FAITH AND WISDOM IN SCIENCE

FAITH AND WISDOM IN SCIENCE

Tom McLeish

Great Clarendon Street, Oxford, OX2 6DP,
United Kingdom

Oxford University Press is a department of the University of Oxford.
It furthers the University's objective of excellence in research, scholarship,
and education by publishing worldwide. Oxford is a registered trade mark of
Oxford University Press in the UK and in certain other countries

First published 2014
First published in paperback 2016

Impression: 1

Published in the United States of America by Oxford University Press
198 Madison Avenue, New York, NY 10016, United States of America

British Library Cataloguing in Publication Data
Data available

Library of Congress Cataloging in Publication Data
Data available

ISBN 978–0–19–870261–0 (Hbk.)
ISBN 978–0–19–875755–9 (Pbk.)

Printed and bound in Great Britain by
Clays Ltd, St Ives plc

To Julie,

Katie, Nicholas, Max and Rosie

Preface

This book has had a long gestation, and I have many people to thank for encouragement, assistance and wisdom along the way. A kind invitation to deliver a keynote address to the 1993 Conference for the Diocese of Ripon, and subsequent discussions with both clergy and lay people in the urban and rural parishes of Yorkshire, first made it apparent to me that the church had need of a theology *of* science, rather than to stand merely a bewildered observer of a battle between theology *and* science. I am especially indebted to the clergy chapter of Ripon for their insights during a retreat devoted to the topic, and to Robin Gamble, Clive Mansell and Matt Pritchard for opportunities to explore its potential role in apologetics and mission. I am grateful to Rolf Heuer of CERN and Richard Burge of Wilton Park for the invitation to an extraordinary multifaith conference on science and religion held under their auspices in 2012. Throughout the journey I have benefitted from the scholarship, critical thinking and unparalleled grasp of theology and science (not to mention friendship, and all those books) of Professor Wilson Poon of Edinburgh University. Some of the ideas presented here have been previously experimented with in articles we have coauthored—they are referenced in footnotes to the text.

During my time at the University of Leeds and subsequently, I benefitted greatly from discussions with Jacquie Stewart, Geoff Cantor, Greg Radick and Simon Robinson. The Leeds History and Philosophy of Science seminar series permitted, among other things, James Ginther to plant the seeds of my interest in Robert Grosseteste, since watered at the Durham Institute for Medieval and Early Modern Studies.

I cannot imagine a more fertile environment than Durham University for any project that requires interdisciplinary advice from generous and scholarly theologians, scientists, sociologists, medievalists, anthropologists and human geographers. I am particularly indebted for comments, critical reading and discussion to: Sarah Banks, Alex Bentley, Richard Bower, Richard Briggs, Giles Gasper, Seth Kunin, Phil Macnaghten, Robert Song, Brian Tanner and David Wilkinson. Welcoming the Durham meeting of the Society for Old Testament Studies gave me the opportunity to meet David Clines, who has generously permitted me to quote from his scholarly translation of Job.

Others who have helped shape the work, commented on the manuscript, encouraged and contributed in many important ways include Denis Alexander, Bill Bryson, Nancy Cartwright, Sarah Coakley, Simon Conway-Morris, Celia Deane-Drummond, Greti Dinkova-Bruun, Sarah Harris, John Hedley-Brooke, Shaun Henson, Terri Jordan, Richard Kidd, Suneel Kunamaneni, Ron Larson, Neil Lewis, Chris Macosko, Julie McLeish, Sue McLeish, Terry Munro, Martin Nowak, Cecilia Panti, David Paton-Williams, Andrew Pinsent, John Polkinghorne, Juanita Rothman, John Pritchard, David Thompson, Faith Wallis, Bryan Wynne, Hannah Smithson and Robert McLeish. I am indebted to Sonke Adlung of Oxford University Press for his patience in keeping the idea for the book 'on file' for longer than it deserved, and for his colleagues' efficiency in bringing it to light. All remaining textual sins of commission and omission are very definitely mine.

The book would not have been written without the patience and the gifts of support and encouragement of my wife Julie and our children, to whom it is thankfully dedicated.

Durham
Feast of St Thomas
2013

Contents

Introduction

Imagine a world, thankfully not our world, without any music. Well, in this sad, dull nightmare, something like music survives in a technical sense. There are, for example, experts in universities and conservatoires in this world who pursue at a professional level a rarefied academic subject called 'sonology' (the word 'music' went out of circulation about two centuries ago). Sonological research, according to government policy, is essential for the economic prosperity of the nation within an international community where sonological status is an important element in national prestige and trade negotiations. The skilled sonologists who carry it out would be the odd ones who excelled at high-school examinations in the difficult subject, and went on to teach it at higher level. They tended to become increasingly good at it by employing all the time in study and practice that their 'normal' colleagues spent playing sport and socialising. Oddly almost no members of parliament, or of the civil service—those responsible for government policy on sonology—actually studied it at a level beyond high-school themselves.

In addition to teaching students, some sonologists pursue incomprehensible research into the playing of a specialist instrument. Others produce manuscripts in the undecipherable hieroglyphs (undecipherable to ordinary people, that is) of sonological notation. These, the experts assure us, encode real pieces of sonological composition that the highly trained experts' ears can make sense of at the academic conferences at which the rare performances are given.

Unskilled members of the public do not attend these performances. They do not have the technical background for it. There are some very good documentary programmes on television, however, that have a significant if small public following. These try to give some sense of what the compositions mean to the experts, explained in simple language and sometimes even playing the tune (when there is one), but leaving out the difficult harmony and counterpoint. There is some debate about whether playing just tunes on their own gives a misleading impression of real sonology, even a dangerously simplistic one. There is a school of thought that prefers to explain just the rhythm, but the viewing figures indicate that the level of arithmetic required to follow rhythm programmes puts people off. Tunes, misleading and oversimple though they are, do better. A highbrow public channel once tried broadcasting an entire movement of a full string quartet, but almost no one tuned in.

Current affairs and cultural programming take a different approach. Sonological items never compete with international or national politics, or even 'comprehensible' art forms such as theatre or film. But towards the end of an hour's serious radio or television, an interview with a sonologist can, it turns out, lighten the tone, providing no great technical concentration is demanded of viewers or listeners. The interviewer, fresh from a deep exploration of boundary rights on the Arctic Ocean floor, seems relieved to take on the easier task of dissuading the dear professor from humming a melody. The essential thing is to be reassured of the economic value to the nation of sonological research—not to be forced to listen to it.

Of course it was not always so in this strange world. Centuries ago music (as it was then called) was, as far as we can tell, a universal good enjoyed by villager, feudal lord and cloistered monk alike. Perhaps in more primitive form, but certainly pipes, drums, strings and whistles indicate musical practice in abundance from the Middle Ages into the Early Modern period. The church was a particularly strong supporter of music and musicians. The discovery of harmony of richer and richer kinds, the setting of ancient texts to music, and the writing of new ones, was central to the worship of the Christian church, and to other faiths as well. The psalms, prophets, history and wisdom books of the Old Testament of the Bible, and the records of the early church in the New Testament, all openly attest to the use of music in praise of God.

So it comes as something of a shock to find that, going into a church today, one finds no music at all. Use of it, then even talk of it, seems to have faded away during the nineteenth century (there had previously been a brief flirtation with the idea of an official ecclesiastical rejection of music during the English Commonwealth). Sonology seems to have become increasingly aligned with a philosophy that insisted on its incompatibility with religious faith and belief. For if humans are able now to create all the sounds that advanced instruments can make, and even write down all possible unheard sounds in sonological notation, what room could there be for a foolish belief that some sort of God was their creator? Not a few of the 'chatting' sort of current affairs media have learned that debates on 'sonology against religion' is a pretty safe bet for listening figures. It is very argumentative, fun to listen to entrenched positions and, as no one really knows much about either any more, fairly safe territory to broadcast on without any difficult homework.

There are a few voices, not much heard, that point out that all this is very sad—that empty concert halls, forgotten music radio stations, youth orchestras as a history book phenomenon mean that people are missing out on a rich cultural experience. Extreme opinions have it that anyone can enjoy sonology. A further strange claim is that one element of the community that might help here are the churches, if only they could rediscover the theological roots of the music they once breathed life into, and received life from. If only they might grasp the fundamentally healing religious purpose of sonology, the glorious sound of a full-voiced choir, the old partnership of Faith and Wisdom in its very foundations. If only. . . .

It is high time to wake up from this nightmarish parable. A world where such a deeply human art as music is marginalised and desiccated to this degree is a profoundly sad one, even to contemplate. But readers will have guessed the topic of the parable. This is indeed our own world—the difference is that with us the problem is not with music but with science. If that sounds strange, then it does so for the same reasons that the suggestion that something was badly wrong with 'sonology' would sound equally odd within that miserable imaginary world.

Yet I want to suggest that science suffers from the same maladies. If it has been marginalised to the orbit of the nerdy expert (or to the media celebrity), if the subtle relationship between science and technology is still widely misunderstood, if urgent public debates on climate, food and disease are repeatedly twisted by misunderstanding of scientific knowledge, if there is no perception that in science there might be sources for contemplation, celebration or culture, then these are signs that we have forgotten the deep, human roots of our minds' search for a bridge into the natural world. We have, simply, forgotten its *story*, and how that story is caught up in the larger narratives of joy, pain, hope, faith and wisdom that make up our religious traditions and cultural heritage.

To unpick and explore the lost currents of faith and wisdom in science it will do no good to jump straight into any form of current 'science and religion' debate (any more than this would have worked in the parable world by looking up 'sonology' in that world's Bible—we, too, are troubled by the words we use). Instead we begin, in Chapter 1, just by listening to the debate's current ebb and flow, the shrill and not so shrill voices. A brief historical survey reminds us of the post-war optimism in an instrumentalist view of science, and the warnings of a nineteenth-century romantic fear of its potential destructive and dehumanising influence. Like the music-starved citizens on their own journey of rediscovery, we need to dig deep into an encounter with the human experiences of science and faith. Only then might we begin to discover what science is *for*.

So in Chapter 2 we try to get around the problems of the troubled associations of 'science' by exploring its older name: 'natural philosophy'. Then we take a journey backwards in time through examples of people 'loving wisdom about nature'. Modern, medieval and patristic examples of doing science by that older name suggest a close look at the biblical material with which their authors would have been familiar. This begins in Chapter 3, but, rather than start with Genesis, we survey the older creation stories of the 'wisdom' literature, and trace some common patterns in their structure and content.

For a scientist, the text that speaks loudest and resonates deepest is found, again not in the contested creation story of Genesis, but in the enigmatic 'book of Job'. So profound is this wisdom book, and so little studied in popular literature, that we spend a whole chapter on it (Chapter 5). Job deals honestly with the chaotic and disorderly phenomena in our world. So we prepare ourselves for the subject of chaos and incomprehension by a look, in Chapter 4, at some current themes in science that link to 'the storm' and 'the earthquake'.

In Chapter 6 our journey through the biblical material on the human relationship with physical creation takes us through the very much shorter time span of the New Testament record. Here, too, we find an infused creation theme, as well as the pain of its troubledness, and visit the only explicit reference in the Bible to the material world without a context of suffering.

Chapter 7 draws together the threads we now hold in our hands: the noise of our public debate, the long history of thinking about the physical world even before this was called 'science' and the biblical material itself.

It outlines what a theology *of* science might look like. Finally some practical consequences are teased out in Chapter 8—with some surprises. Doing this sort of 'grass-roots theology' of science turns out not only to open up some creative solutions to our public tangles over science, but also to suggest radical ways of widely embracing its human, poetic, even holy, richness.

It may come as a surprise that the central chapters focus on the science of ignorance and chaos, rather than on cosmology or the mysteries of life, and on the biblical 'wisdom book' of Job, rather than the more familiar ground of Genesis. But people have always asked themselves seemingly unassailable questions—about the origin of the oceans, what animates living things, what causes change, why the stars shine. . . . I think that this is why scientists are so struck by the torrent of questions about the physical world in the 'Lord's answer' to Job. For one thing, finding the right questions in science, and the faith (there is no other word) that we are able to investigate them, matters so very much more than just finding answers. And *these* ancient questions in Job are full of the chaotic, unpredictable and wild elements of nature. They seem to tease our ignorance at first reading, but then increasingly to hold out the promise of a *story*, a future in which we might understand more than we do now, when our ignorance and fear will be replaced by knowledge and wisdom to use it. *'But where'*, cries a deep voice at the heart of the book of Job, *'where can Wisdom be found?'*

This book is part of one scientist's search for an answer to that question.

1

A Clamour of Voices

What comes to your mind when you hear the word 'science'? I have played this word-association game with groups of all sorts from school lunchtime societies to local women's groups and working men's clubs. I ask for any images, associations or other words conjured up by the idea of science (while requesting that people suppress any unfit for public consumption). Reactions are revealing: 'experiment', 'proof', 'difficult', 'boring', 'test-tube' and 'mad scientist' have all cropped up. If the group does not have any special interest in science the immediate connections are often negative or impersonal. They can also be threatening. Confused and contradictory attitudes often come to light: 'curing cancer' is as likely to emerge as 'atom bomb'. Applying the same game with 'science' replaced this time by 'music', 'art' or 'song' elicits very different responses—usually of individual people or personal experiences. The connections are closer, more coherent, more personal and easier to share. Other seed-words tend also to fall into these two classes: (i) those that generate a very mixed response with a clamour of different voices including the critical, the impersonal and hard-edged, on one hand, and (ii) those that attract warm and more coherent associations, tending more towards personal experience. We might call the first class of ideas 'hard' and the second 'soft'. 'Love' reveals itself as a soft idea almost everywhere, yet, surprisingly, 'Faith' is another example of a hard idea. Like 'science', it flips contradictory switches: 'trust' and 'belief' get mixed with 'blind faith', 'religion', 'extremism'. It triggers signals for widely opposing and highly charged themes in the minds of the group, from peaceful contemplation to fundamentalist-incited violence.

On a much larger canvas, the media both reflects and feeds (perhaps it also exploits) our mixed reactions to hard and soft ideas. Recently an annual television award for science programming was withheld by the judging panel because it deemed no programme produced that year to be of sufficient quality to merit it. The 'entertainment' value had been increasingly perceived by producers to lie in the projection of science as a threat, raising fears and playing to mistrust, rather than engaging viewers in the process of exploring ideas or discoveries. 'Can we trust the scientists?' is a common theme in panel discussions or documentaries. It is a vital question—new infectious diseases and their treatments, our increasingly powerful control of genes in plants, animals and humans, the effect of technological growth on the Earth's climate—these concern us all. Science has become political. To combine two of the 'hard' ideas from the word-association game, 'faith in science', in the first straightforward sense of trusting the investment of people and resources to scientific research was particularly strong in the post-war decade. Here is Pandit Nehru, first prime minister of India:

> *It is science alone that can solve the problems of hunger and poverty, of insanitation and illiteracy, of superstition and deadening custom and tradition, of vast resources running to waste, of a rich country inhabited by starving people.*[1]

In the half-century since then we have learned that science may achieve astonishing things, but arguably without political will it is powerless on its own to achieve Nehru's dream. There is no technical reason why anyone in the world today should starve, go blind with cataracts, have no access to clean water or die of the many preventable diseases. Although science has delivered the knowledge to solve these problems, the wisdom to use it does not seem to come with the package. Even in 'developed' countries there is still no settled political sense of what science is for. In the UK there was no stated political definition of the reasons for state funding of research until the White Paper of 1993:

> *The mission of each research council has been changed to meet the needs of users and to support wealth creation . . . thereby enhancing the United Kingdom's competitiveness and quality of life.*[2]

We will see that this definition does little justice to the historical story of science. Very few of the great scientific discoveries would, at the time, have cleared the bar of meeting 'the needs of users' or supporting 'wealth creation'. The White Paper mission statement also implies a very common confusion of the *two* very different activities

[1] Jawaharlal Nehru, quoted in Atma Ram, The making of optical glass in India: its lessons for industrial development, *Proceedings of the National Institute of Sciences of India* 1961, **27**, 564–5.

[2] Chancellor of the Duchy of Lancaster. *Realising our Potential: A Strategy for Science, Engineering and Technology*. Cm2250. London: HMSO, 1993.

of science and *technology*. But it does project onto the public screen an idolised view of science—in the sense that it is our idols that we perceive will deliver our ideals. Successive administrations in the UK have built on, but not diminished, the political identification of material wealth with the chief justification of doing science. The pattern is similar in other developed nations with dedicated research budgets. Although there is a quiet and downplayed recognition of some value in knowledge for its own sake, even 'pure' science projects such as the particle physics laboratory CERN outside Geneva, Switzerland, are justified first in terms of their technological spin-outs, even when it discovers the long-sought 'Higgs boson'. The logical tensions are buried.

The last decade has witnessed at least a desire to engage a wider public in the political debate around science. Funded initiatives with titles such as 'Public Engagement with Science' by the research councils, and cross-over exhibitions like the Wellcome Foundation's 'Sci-Art' are indicators of deep concerns that a public disengaged with science is politically dangerous. But can we sustain a reasoned and constructive debate when science itself appears on our public stage only in the alternate guises of entertainer and villain? Just listen to the change in voice-tone and attitude of Radio 4 current affairs presenters when they turn from interviewing a government minister to a 'filler' item with an astronomer on the discovery of a new planet. Radio producer and arts graduate Angela Tilby wrote in her book *Science and the Soul* from the perspective of someone with no background in science, yet attuned to the way it is projected:

> *Like priests in a former age, [scientists] seem to guard the key to knowledge, to have access to transcendent truths which the rest of us could never hope to understand. Many people feel that what they do is cut off from everyday life, that it is irrelevant and rather frightening, a form of magic.*[3]

The religious language Tilby employs here is all the more intriguing because the word-association game is played out on the larger canvas of the media when 'faith' is the subject. We are carried by the multiple meanings of 'faith' from the consideration of trust and reliability to all the religious connotations of the word. But in the media this happens on the media's terms. Community service, contemporary theology and soup-kitchens do not sell newspaper copy, web-page hits or air-time.

[3] Angela Tilby, *Science and the Soul*. London: SPCK, 1992.

The threat of extremism, the debate on homosexuality in the church, the social and educational demands of religious groups and the politicisation of religious belief in the crucial power of the USA are all projected and amplified to much higher levels wherever we tune in.

Small wonder then that the noise of confused debate reaches new heights when the two 'hard ideas' of faith and science are brought together, especially when faith now carries its other, religious sense. The debate has become highly political in the sense that quite incompatible positions are held and advocated by people of ostensibly equal authority. One of the most astonishing scientific achievements of the early twenty-first century must be the Human Genome Project—the complete mapping of the DNA code of a human individual that opens a new era of understanding and of medicine. Nearly every one of the trillions of cells making up our bodies contains at its heart a copy of a string-like molecule (this is the DNA) about 2 m long but wrapped tightly into a space 10 000 times smaller than this full stop. Astonishingly, each single molecular copy contains coded in the sequence of chemical 'letters' of the string—3 billion of them—the entire instruction set for the development and functioning of the individual. We have known this for over 50 years now. We have also known some of the possible consequences for new ways of doing medicine if the code were ever read, but the enormous task of extracting and cataloguing it was an unrealisable dream until the 1990s. By then technical advances motivated the launch of both public and privately funded projects to read every letter of the sequence of human DNA.

This 'Human Genome Project' became a controversial race between two teams, generating high publicity and raising a host of issues that questioned once more the faith we place in science and scientists. Should data describing most intimately the molecular detail of *us* become commercial secrets, would science be putting profit before benefit to humankind? The project eventually became a constructive collaboration between a private consortium led by Craig Venter and the publicly funded project directed by Francis Collins. It would be hard to suggest two individuals more qualified to comment on what today's frontier science means for human beings. The journalistic activity circulating around the Human Genome Project, its leaders and its questions of 'faith in science' never lost the opportunity of exploiting the easy linguistic leap to questions of (religious) 'Faith in science'. Perhaps this was made more natural by the sense that the human genome was somehow 'holy ground'—the secret 'words' that bring life to light.

When interviewed in 2007 for a documentary on the future of genetic medicine, Venter was asked about religious belief: 'I do not think that you can be a true scientist and believe in supernatural explanations' was his considered yet clear response, firmly inviting no further question or offering any further comment. Yet Francis Collins has been quite open about his committed Christian belief: 'I can testify that coming to a knowledge of God's love and grace is empowering, not constraining', he wrote in his reasoned book explaining the faith of a scientist, *The Language of God*.[4] Collins sees no contradictions between his essentially traditional form of Christian faith and his equally full-blown scientific world-view. We are left bewildered that two equally informed and intelligent people can express views that at first sight strike us as irreconcilable. Is there an explanation for how these two views can coexist? How can we make up our minds which, if either, is nearer the truth?

Worse is to come, for Venter and Collins represent two of the more moderate voices in the escalating row over science and faith. The zoologist Richard Dawkins of the University of Oxford in the UK has been an outspoken critic of religion and a strong advocate of scientifically motivated atheism for over 30 years. His book *The God Delusion*[5] has poured considerable quantities of carefully extracted oil onto the flames of the argument. His message is that religious faith closes minds, in direct opposition to the open enquiry of science:

> *I am against religion because it teaches us to be satisfied with not understanding the world.*

His colleague in Oxford's chemistry department, Peter Atkins, also a prolific writer of books on science for the general public, goes further. He advocates a view in which science trumps other ways of knowing:

> *Humanity should accept that science has eliminated the justification for believing in cosmic purpose, . . . science can illuminate moral and spiritual questions.*[6]

Like Venter, these two voices see religious faith directly competing with science while buying into a view of science itself that exalts it into a unique channel of knowledge about the world (some writers call this an 'epistemology'), including the world of minds and purpose. This claim

[4] Francis Collins, *The Language of God*. New York: Free Press, 2006.

[5] Richard Dawkins, *The God Delusion*. London: Bantam, 2006.

[6] Peter Atkins, Will science ever fail?, *New Scientist* 1992, **8 August**, 32–5.

is part of the reason for Angela Tilby's fear, implicitly comparing today's scientists with former 'gnostic' communities whose priesthoods claimed secret and exclusive routes to truth.

Of course Dawkins, Atkins and others who propose an exclusive role for science as our way of knowing find many equally extreme positions to tilt against within fundamentalist wings of religious traditions. It is one of the great social surprises of the twentieth century that, amid those very nations at the front of the enormous rise of knowledge of our universe's workings and their technological consequences, extremely doctrinaire forms of belief have grown rather than receded. Particularly in the USA there are many relatively mainstream Christian churches in which members are expected to reject an ancient origin for the world, and the gradual evolution of life forms on our planet, for a literal interpretation of one of the (many) creation stories in the Bible. Of course insistence that humans appeared suddenly a few days after the origin of the universe itself makes it impossible for anyone informed about the way we have arrived at our present view of the cosmos, let alone professional scientists, to take this view seriously. The fact that many manage to struggle on in churches like this reminds us of humankind's dangerous ability to live within contradictions. But it is not a happy existence, and it creates easy targets for the invective of those who see science and religion as mutually irreconcilable.

It is instructive for a European readership to see the extreme care that American scientists who are also believers feel they must take when writing for their congregations. Francis Collins's book, which we glimpsed into above, contains an extended and gentle argument for evolution, treading so carefully through the topic that we suspect that he fears that at any moment his book will be hurled against the nearest wall by his reader, whether they be fundamentalist Christian or atheist. Listen carefully behind the words of another Christian biologist, Darrel Falk, who has written to his evangelical colleagues as one of their number about 'coming to peace with biology':

> *Let us not allow a particular interpretation of a tiny section of God's precious Word to become so central that it creates a gulf blocking the access of any individuals to the experience of God's love in the church. I almost missed out, so wide did the distance seem to me.*[7]

[7] Darrel R. Falk, *Coming to Peace with Science*. Downers Grove, IL: Intervarsity Press, 2004.

It is worth reminding ourselves that attempts to set up science and religion against each other as competing epistemologies are by no means recent phenomena. Thomas Paine, one of the intellectual fathers of the independence movement in eighteenth-century America, inveighed against what he perceived as the intellectual pretensions of religious faith. He chose to attack not the common religious practices of faith-based communities, but the academic field of theology:

> *The study of theology, as it stands in the Christian churches, is the study of nothing; it is founded on nothing; it rests on no principles; it proceeds by no authority; it has no data; it can demonstrate nothing; and it admits of no conclusion.*[8]

The background to a critique worded in terms such as these must of course be a pattern of disciplines that *do* study 'something', rest on founding principles, admit of (expert) authorities, collect data and make demonstrations of their conclusions. Paine did not have to be explicit: this is the pattern of the sciences of physics and chemistry that over the previous century had been successfully discovering mathematical laws of immense generality and power. Newton's theory of gravity described the motion of the planets and their moons with perfect precision; the new molecular ideas of matter had begun, in the hands of Boyle, Priestley and Lavoisier to explain chemistry and the behaviour of gases, liquids and solids. Venter, Dawkins and Atkins inherit a line of reasoning that gives the period that saw the rise of modern experimental science its name: the 'Enlightenment'—for what is illuminated if not darkness? Dawkins stands in two centuries of tradition when he echoes Paine:

> *What has 'theology' ever said that is of the smallest use to anybody? When has 'theology' ever said anything that is demonstrably true and is not obvious? What makes you think that 'theology' is a subject at all?*[9]

Whatever we might think of the benefits of theology, it is the implication behind this narrative that surfaces in Paine and Dawkins—that science is *the* 'subject', *the* route to all knowledge—that gives people like Tilby cause for concern. There has been a palpable move from an uneasy coexistence of science and theology to this, more aggressive and public stance in the last decade—writers who align with it have even acquired

[8] Thomas Paine, *Age of Reason*, Part II, Section 21. Paris: Barrois, 1795.

[9] Richard Dawkins, The emptiness of theology. *Free Inquiry magazine*, 1998, **18**, no. 2.

a label of the 'new atheists'[10] (others often grouped under this heading include Sam Harris, Daniel Dennett and Christopher Hitchens). Reading the corpus of these authors since 2001, it is hard to avoid the conclusion that much of the anger pushing the paragraphs along has been released by the '9/11' terrorist attacks on New York and Washington by Islamic extremists, and subsequent events. It is certainly true that new-atheistic arguments focus increasingly on the actions and statements of extreme (and in most cases unrepresentative) elements of religious faiths. Where Dawkins and Dennett work on the storyline that has science and religious faith at irreconcilable war, Harris and Hitchens focus on the political, cultural and psychological as a starting point:

> *There still remain four irreducible objections to religious faith: that it wholly misrepresents the origins of man and the cosmos, that because of this original error it manages to combine the maximum of servility with the maximum of solipsism, that it is both the result and the cause of dangerous sexual repression, and that it is ultimately grounded on wish-thinking.*[11]

Hitchens is in no doubt that religion is a moral evil, but his evidence—although powerful—is selective and anecdotal. Dennett launches, or re-launches, a call for an ambitious scientific project: thoroughgoing, evidence-based scientific evaluation of the benefits, or otherwise, of religious faith (there have been a number of selective psychological and statistical studies published on, for example, the lifespan of those with or without religious belief). He wants more—to understand religion using all the light that anthropology, psychology, neurology and sociology can throw upon it:

> *So here is the prescription I will make categorically and without reservation: Do more research. . . . My task was to demonstrate that there was enough reason to question the tradition of faith so that you could not in good conscience turn your back on the available or discoverable relevant facts.*[12]

This is intriguing—taking the bombast and rhetoric out of the conversation, Dennett coolly reminds us that there is no such thing as a 'boundary between science and religion', at least from the viewpoint of

[10] Victor J. Stenger, *The New Atheists: Taking a Stand for Science and Reason*. New York: Prometheus NY, 2009.

[11] Christopher Hitchins, *God is not Great*. London: Atlantic Books, 2007.

[12] Daniel C. Dennett, *Breaking the Spell: Religion as a Natural Phenomenon*. London: Allen Lane, 2006.

science, for its domain of explanation is no less than all that exists. We will have more to say on this later, especially on the question of whether one can support more than one epistemological framework that takes everything as its object. If science can illuminate everything, must it do so exclusively? This is certainly the 'scientistic' view of the new atheists and their movement.

There have been strong reactions to such claims that science is the unique route to knowledge. A persistent voice of caution comes from historians of thought, who remind us that behind all of our current assumptions there is a narrative—a story of ideas running through centuries and across languages and nations. Furthermore, the currents of narrative run in both directions—while our own thinking is influenced by the generations before, so we tend to colour our accounts of history retrospectively. We tell stories of the past in defence of our present positions. So an account of the seventeenth-century 'Enlightenment' from the perspective of twenty-first-century atheism tells a tale of a new and entirely secular empiricism banishing a theologically incarcerated scholasticism left over from the Middle Ages. Science becomes in this story a secular project from the outset—the laws of physics removing the need, for example, for the guiding hand of a Creator. John Hedley Brooke's persistent reminders of the historical evidence tell, however, a different story. He points out, for example, that the emergent mechanistic world-view became, for the seventeen-century thinker Mersenne and others, strong supporting evidence for divine cosmic carefulness.[13]

Furthermore, not a few of the central Enlightenment scientists were explicit in their framings of theological motivations for science. Johannes Kepler, Robert Boyle and Francis Bacon all in different ways saw their projects as answering a Christian call to light up, or even restore, the world. Peter Harrison[14] has argued strongly that the founders of the Royal Society perceived their task as winding back the darkening ignorance brought about by the 'Fall' (of Adam and Eve). He espouses a radical rethinking of the shift from medieval to early modern thought in regard to science:

[13] John Hedley Brooke, *Science and Religion: Some Historical Perspectives*. Cambridge: Cambridge University Press, 1991.

[14] Peter Harrison, *The Fall of Man and the Foundations of Science*. Cambridge: Cambridge University Press, 2007.

> *Indeed, surprising as it may seem, what distinguishes seventeenth century discussions on knowledge from scholasticism is not their secular character but rather the fact that they tend to be more explicit in their reliance on the resources of revealed theology than their medieval equivalents.*

So Collins and Falk have historical precedence within modern science for their insistence that a religious commitment raises no necessary conflict with a scientific search for knowledge.

Interestingly the loudest voices in protest against scientistic approaches to knowledge come not from the religious traditions chiefly under attack, but from a much wider circle. Journalist Brian Appleyard sees great danger along the road of scientism:

> *The heartless truths of science . . . have depersonalized . . . dehumanized . . . , we need a humbling of science.*[15]

Voices like this—critical of science today—stand in just as long a tradition as those who would idolise it. The Romantic poets of the nineteenth century resonate with Appleyard's dismay. Take John Keats for example (in *Lamia*):

> *Do not all charms fly*
> *At the mere touch of cold philosophy?*
> *There was an awful rainbow once in heaven:*
> *We know her woof, her texture; she is given*
> *In the dull catalogue of common things.*
> *Philosophy will clip an angel's wings,*
> *Conquer all mysteries by rule and line,*
> *Empty the haunted air, and gnomed mine*
> *Unweave a rainbow.*

Here 'philosophy' stands for the ideas we would now label with 'science'—but we see clearly what impression they made on Keats. His world contains two classes of ideas: 'charms', which excite, delight and cause to wonder, whereas 'common things', which contain no mysteries for us, are all explained, all dull and all mathematised or geometrised. For Keats, science relentlessly moves everything from the first class to the second, it saps nature of all that touches the deeply human within

[15] Bryan Appleyard, *Understanding the Present: Science and the Soul of Modern Man*. New York: Doubleday, 1992.

us. He is even nostalgic for the ancient fears of ghosts that science dispels (he need not have worried, but was born more than a century too soon to be comforted by *The X-files*). Even though the existence of the rainbow, unlike that of ghosts or gnomes, is in no way doubted by a scientific explanation,[16] nevertheless that very mode of approach to the beautiful atmospheric phenomenon of refraction represents an 'unweaving' of a beautiful tapestry so that we can no more feel its texture. Only the threads that once made it up lie in a heap upon the floor. It is a dismal metaphor for science.

In visual form the romantic loathing of science as the great desiccator of art and imagination famously finds its greatest expression in William Blake's painting of Newton bent double over a pair of dividers *at the bottom of the sea.*[17] So consuming is the scientist's act of measuring the dull details of the sandy floor that he never spares a moment to gaze upon the beauty and lustre all around. In more modern setting, the accusation that, in contrast to art, science somehow dehumanises us finds voice everywhere. We have already heard from Brian Appleyard. The artist Georges Braque is more specific:

> *Art is made to disturb. Science reassures. There is only one valuable thing in art: the thing you cannot explain.*[18]

Sometimes I am given the enjoyable opportunity to discuss with high-school groups aspects of their 'general studies' courses in the year before going to university. The differences between the arts and the sciences is a favourite topic; the brighter students who did not choose to study science at that level often explain that they felt that science offered no room for imagination or creativity—that scientific explanation deadens. The objection to science, let alone scientism, that it crushes the spirit of human creativity, and which has no apparent religious source at all, is very much still in circulation.

I do not want to address or defend any of these views at present, just to hear them. We will need to spend some time exploring what science is

[16] Explanations of the rainbow possess a fascinating very ancient history beginning at the latest with Aristotle. The identification of refraction as the key underlying phenomenon was first made by Robert Grosseteste (*De Iride* c.1220), whom we shall meet in the next chapter, and an essentially correct explanation at the classical level was offered by Theodoric of Freiburg (*De Iride* c.1310).

[17] Blake's 'Newton' is on display at the Tate Britain, and in electronic form at https://www.tate.org.uk/art/artworks/blake-newton-n05058.

[18] Georges Braque, *Le Jour et la Nuit: Cahiers 1917–52*. Paris: Gallimard, 1988.

and what it is not before we can separate truth from impression, or ancient narrative from reasoned argument. Especially within the school educational setting, personal experience weighs heavily. When years spent not arriving at the answers in the back of a worn textbook can be contrasted with the fun to be had in a lively drama studio or mixed-media art class, who could blame a young Keats or Blake? The delightful, if frightening, confrontation with 'the thing you cannot explain' is a deeply human desire—but already those with some experience of the sciences might be wondering why someone like Braque has not realised that they, too, can be paths to this confrontation with the alien and unknown, and as such can be far from reassuring. In a remarkable book *Science, the Glorious Entertainer*, written in the 1960s, but not so well known now, Jacques Barzun touches a nerve that may lead to the source of science's apparent failure to deliver the confrontation with 'otherness' and the world of the imagination that the arts seem to be able to do. 'Science is not with us an object of contemplation' is his telling observation.[19] Perhaps Barzun is right—could I really stand absorbed and focused for half an hour gazing at the idea of an electron? It begins to become clearer why scientism draws such revulsion from so wide a circle of opinion: in spite of its attempt to direct its ammunition at the idea of faith in God, it has been systematically perceived to be tilting at the idea of faith in humanity.

It is worth reminding ourselves that the tension between science and humanities is by no means confined to books, lecture halls and talk-channel radio. The problem of the impact of science on what it means to be human is currently hotly debated in the relatively new arena of public committees dedicated to ethical and regulatory research practice in medicine, agriculture and nanotechnology. It is not that the science is new, but that when science is applied to the deepest structures within human beings we find ourselves for the first time playing the double role of observers/manipulators and observed/altered. Our current confusion over where and what is 'human' has caused some unexpected decisions. A telling example is cited by Derek Burke, chairman of the UK's Advisory Committee on Novel Foods and Processes (1988–97). His committee was asked to evaluate the public acceptability of a technology to harvest drugs from the milk of genetically modified sheep. A section of DNA that codes for the desired molecule is inserted into the entire DNA of the animal in such a place that the drug would then be

[19] Jacques Barzun, *Science the Glorious Entertainment*. New York: Harper and Row, 1964.

expressed in the animal's milk, from where it could be extracted straightforwardly. After the usual process of wide consultation a problem was found—not with the genetic modification as such, but because of the necessity that most such animals would also need to be farmed for meat for the process to be economically sustainable. For the new section of DNA is a copy of a tiny piece of the *human* genome. The committee concluded that a public abhorrence of eating anything that appears traceably human prevented adoption of the entire technology. Unofficially and euphemistically called the 'yuk factor', we find ourselves here coming into contact with a dark stream of fear that goes well beyond the protests of Keats.

Few would consider his friend Mary Shelley's novel *Frankenstein* within the first rank of English literary works, but for giving voice to a deep suspicion of science and for providing an icon to the horrors that might result from turning science onto ourselves, it is matchless. Jon Turney has charted the development of the idea in the century and a half since its creation in his *Frankenstein's Footsteps*.[20] He shows that the monster and its creator came rapidly not only to reflect a public view of science and scientists, but also to frame it for all subsequent generations (recall the 'mad scientist' response of our word-association game). The idea translates very easily between media, and tarnishes by association without debate—think of the opinion we are tempted to form automatically by reading the single tabloid heading 'Frankenstein foods'. The novel also placed implicitly before the public eye an opposition between those who write (and read) literature and those who experiment and theorise about fact. Most especially it threw into moral doubt those who crossed beyond the threshold of measuring nature to changing it.

Some of these confusing literary and social currents that swirl around the idea of science in the public mind resurface, even unconsciously, within a debate now two generations old in the English-speaking world. The idea that our society is structured around 'two cultures' (the artistic and the scientific) was brought into focus in a famous lecture and book by C.P. Snow.[21] Comparing the cultural standing of the works of Shakespeare with the second law of thermodynamics, Snow famously railed against the requirement of a 'cultured' person to know the first

[20] Jon Turney, *Frankenstein's Footsteps*. New Haven, CT: Yale University Press, 1998.

[21] C.P. Snow, *The Two Cultures*. Cambridge: Cambridge University Press, 1998.

but not the second. The claim and the responses of the ensuing public debate, though they now look shallow, have left the legacy of a combative style that we now have to work harder to overcome. A strong echo of the debate resounded around the bitter exchange of books and articles in mainly academic circles during the 1990s that became known as the 'science wars'. Unleashed by the publication of a spoof article in a sociology journal written by physicist Alan Sokal, who was disenchanted with what he saw as meaningless pretensions of some disciplines within the humanities, it drew back the curtain on a clearly territorial battle on truth within our universities. Interestingly it also ignited fires in the French-speaking world (Sokal's later book *Impostures Intellectuelles*[22] was published first in French before an English translation appeared). To many scientists it was a shock to see to what extent they had failed to win any recognition from their academic colleagues in literature and philosophy—that their own 'faith in science' as a route to knowledge of the world was by no means an assumption shared universally across late twentieth-century campuses.

A deeper exploration of the fault-lines between scientific and artistic disciplines has been pursued over many years by the critic and writer George Steiner. One of his paradoxical criticisms of the arts is that they pretend to be sciences. Artists and writers have been mesmerised by the success of scientific research, and too keen to adopt in their academic communities the same customs and structures—this for Steiner turns our universities into 'secondary cities' where commentaries on commentaries are churned out in place of fresh new art. He imagines a 'primary city' where the only form of criticism allowed of a poem is another poem, of a novel another novel. Yet he claims that in the sciences:

> *Today, it is noon-time not in the arts but in the sciences. An estimated ninety percent of all scientists in history are now alive. Whereas the study of the humanities, the editions of the classics, the performance of established western music, . . . looks backward, science is, by very definition in forward motion. . . . In the theoretical or applied sciences, even a middling talent is on an upward escalator. . . . Theorems will be solved, crucial experiments performed, discoveries made next week and/or the week thereafter.*[23]

[22] Alan Sokal, *Impostures Intellectuelles*. Paris: Éditions Odile Jacob, 1997.

[23] George Steiner, 'A festival overture'. Lecture, University of Edinburgh, August 1996.

Strangely, Steiner reacts to his observations of science in the opposite way to Braque or Blake (or to my disenchanted high-school students). When he looks from one 'culture' to the other he seems to see very green grass in the neighbour's garden—something very like the 'primary city' that he longs to regenerate within the humanities. Oh that it were true! Those of us who work within science know very well that the consequence of many relentless pressures—to publish, to be noticed in a way that contributes to the 'world class' status of our own institutions, to win more and bigger research grants than the university in the next city, not to mention the impoverished state of our own imaginations and insights—give Steiner's words a haunting hollowness. The reason that 'theorems will be solved . . . next week or the week thereafter' is largely that, if they are not, we lose the eternal race to the top. Perhaps it feels a little like the predicament of Scheherazade: every evening inventing another new story for her Sultan, for if ever the stories dry up she loses her head.

This is surely not what 'science is for'—but in Steiner's critique of the arts, and in his articulated longing for what they might do for people if they were set free from the self-referential treadmill, there is a remarkable clue: at one point he seems to look at the strangeness of the physical world around him with new eyes, and becomes almost frightened at the inhumanity—not of any sort of science or scientific process—but of the universe itself. There is a huge gulf between ourselves as living thinking beings and the unspeaking material forms around us. How should we bridge such a great divide? Steiner's answer comes as a surprise after all he has said before about the failure in his eyes of art to find its way: 'Only art can go some way towards making accessible, towards waking into some measure of communicability, the sheer inhuman otherness of matter . . .', he writes in his deeply felt discussion of meaning and language, *Real Presences.*[24] 'Only *art*'? To a scientist this comes as a shock—to me as a wake-up call to think what really motivates us at the deepest level to explore the world. If science is not there to establish lines of communication between our minds and the 'sheer inhuman otherness of matter', then what is it doing? Why does Steiner, so sensitive to our human need for some sort of reconciliation with our world, not see science as part of the answer when it surprises us over and over again with our ability to reveal the patterns beneath the things we see, hear, feel and touch by careful observation,

[24] George Steiner, *Real Presences*. London: Faber, 1989.

imagination and theory? Perhaps in spite of his apparent familiarity with the competition-driven parade of the science community, he speaks for many people when he denies it a role in the inspired contemplation and recreation that he does see in art.

A remarkably similar, though more accessible, reaction to the 'otherness' of the world appears in author Bill Bryson's introduction to his excellent and entertaining history of science *A Short History of Nearly Everything.*[25] He relates the moment when, on a long aeroplane journey, he looked out of the window at some intricate cloud patterns and found himself wondering how and why they formed. This led to a chain of musings on natural phenomena and their causes—and the horrific realisation that he knew nothing about any of these things. He and Steiner use different words, but they speak of the same sort of divorce between ourselves and the world in which we are, paradoxically, so physically embodied. So began a second personal journey for Bryson into science, through extended conversations with people who lead research today in physics, biology, chemistry, astronomy and geology. He became fascinated by the history of science, and above all in a quest for understanding how we know the things we know. His delightfully human account resonates with Einstein's famous surprise that 'the most inexplicable thing about the universe is that it is explicable'.

A friend of mine who graduated in the languages is one of many without any science background who has found Bryson's book a doorway into a new way of feeling a little more at home in the world. This is very telling, because, as its scientific readers will aver, the substantive content of the science is not circumvented. While it may avoid using mathematics, the story does not sacrifice technical depth for the sake of a gripping pace—it manages to weave both together. Is there a clue in the self-realisation of the need to reconcile the human mind with the natural world that might nourish discussions between and within our fragmented world of arts and sciences? And if there is, might the depth at which this structure lies (deep enough to perceive commonalities of cultural purpose between physics and art, for example) also help disentangle some of the noisy debate around science and religion?

Turning once more to Steiner, we find the same concern that our post-modern world is somehow de-humanising us that we saw in the harsh critics of scientism. But in this case we also pick up the strands of

[25] Bill Bryson, *A Short History of Nearly Everything*. London: Transworld, 2003.

someone in sympathy with the excitement of science and who claims at least a ringside seat at its current triumphs. He seems also to be at a safe distance from the confusion with religious faith—a post-holocaust Jew with impeccable humanist credentials. Yet there is a big surprise in store, for Steiner sets his whole discussion of our need for meaning, his fear that language has lost its connection to the physical world, within an explicitly religious referential world. The otherness, or inhumanity of matter, corresponds to a 'broken contract'—we have lost faith in the meaning of language in a way that mirrors our current loss of faith in science. In a 'post-modern' world where my reading of any text is as valid as yours, and carries no assumption about any meaning encoded by a 'writer', whether a writer of a book or a 'Writer' of the physical creation itself, there is no need to risk anything, no need to trust, no need to put our faith in anything outside our personal worlds. His analysis that we have turned away, or are in danger of turning away, from an obligation to establish the ties of meaning between our minds and the world we inhabit sounds very like an Old Testament prophet warning Israel that it has rejected the Covenant Law that was supposed to guarantee it life. A greater shock awaits: for our present uncomfortable state of tension with our world has a history—and that history a shape that Steiner invokes an explicitly Christian image to convey. It is the shape of Easter, the shape of Friday, Saturday and Sunday. It is worth quoting in full his reasons:

> *There is one particular day in Western history about which neither historical record nor myth nor Scripture make report. It is a Saturday. And it has become the longest of days. We know of that Good Friday which Christianity holds to have been that of the Cross. But the non-Christian, the atheist, knows of it as well. . . . We know, ineluctably, of the pain, of the failure of love, of the solitude which are our history and private fate. We also know about Sunday. To the Christian, that day signifies an intimation . . . of resurrection. . . . If we are non-Christians or non-believers, we know of that Sunday in precisely analogous terms . . . the day of liberation from inhumanity and servitude. . . . The lineaments of that Sunday carry the name of hope. . . . But ours is the long day's journey of the Saturday. Between suffering, aloneness, unutterable waste on the one hand and the dream of liberation, of rebirth on the other. In the face . . . of the death of love which is Friday, even the greatest art and poetry are almost helpless. In the Utopia of the Sunday, the aesthetic will, presumably, no longer have logic or necessity. The apprehensions and figurations in the play of metaphysical imagining, in the poem and the music, which tell of pain*

> *and of hope, of the flesh which is said to taste of ash and the spirit which is said to have the savour of fire, are always Sabbatarian. They have risen out of an immensity of waiting which is that of man. Without them, how could we be patient?*[26]

The writer is by no means a signed up member of any church, but at the centre of our questions of loss of faith in the sciences, and of some scientists' rejection of faith altogether, he draws us back to a powerful story of faith—arguably the most powerful of all. The significance of the 'three-day story' is one we can all recognise within our own experiences, small or great. It has the status of 'myth', not in the sense that it is necessarily make-believe but in the definition of J.R.R. Tolkien—a story that we all know without knowing when we heard or read it[27] (I am reminded of C.S. Lewis's shocking description of the New Testament Easter events as the 'true myth'). Friday is the past time of our loss, hurt, failure, unanswered questions. Sunday is the mental heartland of our hope, for healing, fulfilment, answers. But for now we live in the Saturday time of waiting, of irreconcilable positions, of the clash of cultures and the broken contract with nature, of the interminable feeling that we are strangers in the presence of 'the sheer inhuman otherness of matter'. It is the time of faith, whether we are religious or not, for no future is certain, and there is enough trouble to tempt us into despair whether we are writing books that aim to dispel all religious belief as ancient superstition in the dazzling light of science, or equally convinced that the theory of evolution is an evil deception. It does not seem possible to separate the questions of 'faith in science' (faith in the sense of trust) wherever we turn, from 'Faith in science' (faith as religious activity). Even writers like Dawkins and Atkins never extricate themselves from the second question, but draw palpable energy from it. This is hard to understand if it is really true that religion is old, outworn and meaningless, and science new, epistemologically sound and the successful competitor for the same ground that religion has occupied for centuries. The human connection between the two runs too deeply for that.

There is a hint of a rich new narrative behind Steiner's linking of the current work that both arts and scientists do within human culture to the Easter story. It would be exciting if this narrative contained the

[26] George Steiner, *Real Presences*. London: Faber, 1989.

[27] J.R.R. Tolkien, *On Fairy-Stories*. The Tolkien Reader. New York: Ballantine Books, 1966.

potential to resolve some of the angry impasses we have surveyed. But, if it is there, it is surely deeply buried: none of the discussion swirling about the new atheism or the 'science wars' makes reference to a story of the human purpose of science. Perhaps this is because of the notion that science as such is a post-Enlightenment innovation, and so cannot belong to any much longer cultural history. Perhaps it is because we have more than glimpsed that a long human story that successfully resolves where science belongs would inevitably draw on theological tradition. We already know that theology appears embattled, forced into a corner, at least within the publically projected form of its conversation with science. This has not endowed it with a confident voice (except in the case of narrow and confrontational dogmatists)—the idea of constructing a 'theology of science' is an urgently needed, but missing, voice. The new task before us is to explore such a narrative—we will require a journey back to some very old sources within the ancient literature of 'wisdom'. The outlines of a theological story of science will become clear enough to attempt writing them down by Chapter 7.

But there is one great obstacle to hearing clearly any single one of the clamour of voices that make up the chattering crowd present in this chapter. It is that science itself is perceived frequently through the tinted lenses of others, whether of Steiner's noon-day hue or the poisonous shadow of Mary Shelley. We will need to take seriously Jacques Barzun's challenge and take a contemplative journey of our own through some science before we can tease out roles for faith, or even 'Faith and wisdom', in this highly emotive strand of our culture. What does it feel like to do science? We need to experience it through the work of thinkers both today and in past centuries, before we ought to talk much more about its purpose. Some rolling up of sleeves and 'contemplation' of science from within will also prepare us for resonances we will need to be sensitive to in reading wisdom literature from much older cultures.

2

What's in a Name? Stories of Natural Philosophy, Modern and Ancient

Part of the problem is the name itself—why does speaking of 'science' sometimes make people feel uncomfortable? More to the point, why is it that some of the negative associations we heard about in the last chapter seem glued rather firmly to that word, rather than the stories, the communities or the ideas that it represents? I have regularly tried another activity with groups of people—adults and children—that follows the word-association game with which we began. I point out that 'science' and 'scientist' have Latin origins in the verb *scio*—'I know'. Whether we know Latin or not, the complex associations built in our minds by language will have linked science implicitly with a claim to knowledge, even before we read that etymology points towards those connections. But 'scientists' were not always so called: we know that the word was coined around 1830—probably by William Whewell, the polymathematical master of Trinity College, Cambridge. Before then if any collective expression were used for those who made it their business to examine the heavens or to explore the chemical properties of gases or the distribution of different rocks and the varieties of flora and fauna on the Earth, that expression would be 'natural philosopher'. The etymology could not be more different: the older name replaces the Latin *scio* with the two Greek words, *philia* and *sophia*, for 'love' and 'wisdom'. Ask yourself: what happens to our image of science if we replace in our minds its word-label '*I know*' with '*I love wisdom to do with natural things*'? Instead of a triumphal knowledge-claim we have a rather humbler search, together with more than a hint of delight. We also have as a goal something deeper than pure knowledge, in the *wisdom* that surrounds and supports it.

The idea of wisdom draws on a long history of Greek and Hebrew ideas in which '*Sophia*' has been personified to an extent that '*Scientia*' never could be—ancient writers could imagine talking with someone called 'Wisdom' but not someone called 'Knowledge'. Finally, at the heart of 'natural philosophy' there is the word for love. Not nowadays an idea that readily claims association with science, it belonged there once. I have often seen a smile and an 'I wonder . . . ' expression appear on the faces of people who a moment before have claimed to find no interest in the cold, logical inhuman process they imagine science to be, when they begin to think of the new directions in which 'love of wisdom of natural things' might take them. At the very outset there is nothing threatening them anymore, no forced confrontation of their apparent lack of knowledge in the face of the insurmountable demands of a textbook or examination paper. There are no demands for worked answers to be compared with those on the solution sheet, but instead something much more childlike and inviting, a curious searching for questions, the offer of undemanding observation. All this and more is heard and felt afresh.

But is this just a tempting and comforting illusion? Surely the rigorous and competitive search after knowledge is, after all, a better description of what scientists do than 'love of wisdom of natural things'? This is only answerable by a close examination of what happens when people actually do science. As part of our examination of what 'faith and wisdom in science' might mean, we must be sure that we know what science is, rather than rest content with the images projected by its advocates or critics, or even by school textbooks. To do that, we will have to listen to some stories. In this chapter these will not be the usual science stories of the history books, for they are as well known as they are unrepresentative of everyday experience. We will need to visit the library, however, for I want also to take a journey through history, tracing if we can the pattern of 'love of wisdom of natural things' in reverse, from the present day back to more ancient strata within which we can recognise its cultural–evolutionary pathway. So the stories begin with projects and people who I know well in their ups as well as their downs. As we turn the clock back and pay a visit to some extraordinary people who do not commonly crop up in histories of science, as well as some who do, there may be some surprises.

The Case of the Surprising Jelly

I work in a university department of physics in the UK, among a group of mostly young women and men interested in understanding unusual behaviour in fluid or soft matter. To get the general idea, think of the strangely elastic properties of materials like bread dough, the slimy track laid down by slugs and snails, the hot sticky spread of cheese on fresh pizza (and the industrial equivalent of melted plastics on their way to being squeezed into the shape of car bumpers or mobile phone cases), or the optical properties of the liquid crystal displays in your watch.

It is an interesting field, not least because it introduces us to a wide spectrum of very different people who think about science in completely different ways. Some of the fluids we study come from chemists, some from biologists, some from industrial engineers. Our laboratory gives us all sorts of imaginative ways of measuring the properties of the materials, but the real goal of the work is to understand *why* they flow or recoil or change colour in the way that they do. This requires us to go beneath the surface appearances of the fluids, both in measurement and in thought, to perceive what we can of their inner structure and workings. That was why a colleague from the chemistry department once asked for help about a fluid that he and a student had made, but which had turned into a strange jelly. Before I recount what happened next, we had better take a look at the most useful mental tool that we use in science of just about every kind. It is called the 'molecular theory of matter'.

This idea is so powerful and yet so simple that it is worth exploring for a while. One of the great scientists of the twentieth century, American physicist Richard Feynman, who won a Nobel Prize for his work on how light and matter interact, posed himself the question: if allowed only a single sentence, how would he pass on the greatest achievement of civilisation to a new generation trying to rebuild the human race after some destructive cataclysm? His answer was:

> *I believe it is the* atomic hypothesis *(or atomic* fact, *or whatever you wish to call it) that* **all things are made of atoms — little particles that move around in perpetual motion, attracting each other when they are a little distance apart, but repelling upon being squeezed into one another.** *In that one sentence you will see an* enormous *amount of information about the world, if just a little imagination and thinking are applied.*[1]

Think for a moment how this extraordinary idea works to enrich our observations of the most commonplace of experiences. Water is strange

[1] R.P. Feynman, *The Feynman Lectures on Physics*, vol. II. New York: Basic Books, 1964.

stuff: usually it flows easily, filling glasses, jugs, streams and oceans. But cool it and it solidifies into the ice-forms of bergs, snow and sheets. Turn on the kettle and it transforms once more into the springy, space-filling gas we call steam—it once turned locomotive wheels with its pressure. We know that these three forms do not represent permanent changes, for ice melts and steam recondenses. How is it then that matter can change its form, without changing its substance?[2] Thinking with the molecular idea resolves the puzzle in an extraordinarily creative way. If a body of water consists of billions of the smallest units (we call them the water *molecules*) with the properties Feynman described, then all the changes may be attributed to changes in the degree of motion of the molecules and to their spatial configuration, not to their structure or individual nature. At their most rapid, the momentum of their mutual collisions overcomes the attraction they have for each other when close. They pass by each other too rapidly for their attempted hand-grasps to succeed. This is the underlying picture of the gas form. Its pressure comes from the same molecular motion, but now in the form of a myriad continual collisions with the walls of its container.

Cool the collection of whirling particles and they slow enough for the close-range attraction to begin forming clumps, droplets of many molecules. Only a few now escape from the sea of liquid, where the molecules still have sufficient motion to roll around each other but not permanently to escape from their neighbours. Cool once more, however, and the remaining motion is insufficient to mask the special nature of the forces between molecules, whose geometric shapes mean that attractions are stronger in some directions than others. They fall into a sort of scaffold structure in which each links with a few others in stable orientations: the water has frozen into ice. If the slowing down we call 'cooling' is done carefully enough, then the special directions in which the molecules attract each other most powerfully persist for long distances through the ice, and beautiful crystal forms appear. The queen of them all is the snowflake.

What a powerful explanatory idea! But it is also creative because it suggests new questions and new ways of 'looking' at the world or, in this case, specifically at the nature of water. We might start by asking where

[2] This is a very old question indeed. It appears in the West in Heraclitus, Leucippus and Democritus in the fifth century BCE, and in India in the sixth century BCE; see Thomas McEvilley, *The Shape of Ancient Thought: Comparative Studies in Greek and Indian Philosophies*. New York: Allwarth Press, 2002.

the attractive forces come from. Are they connected with other forces that we experience? We know, for example, about the electric attraction that pins children's party balloons to the walls when they are rubbed on a dry sweater. If the source of these two attractions really is the same we might expect water itself to respond to the electric effects of our charged balloon. That is simple to test—just run a thin stream of water from a jug or tap past the charged balloon, instead of sticking it immediately to a wall, and watch closely. Perhaps we should at this point try the experiment rather than read about it[3]—but when we have done so we will have learned something remarkable about our world that will never leave us.

The elastic or slimy fluids that especially interest the research group arise from the special form of their molecules. Rather than the compact forms of those in water, the molecules of a plastic substance, for example, take the form of long flexible strings. The restless motion that produced the phenomenon of pressure in steam now makes the molecular strings dance and twist in a continual shimmering salsa. Sometimes they get tangled in little local knotted regions where the local attractions win out over the heated tendency to move around. These linked regions tie down the ends of the dancing molecular chains so that, although the material as a whole may deform and stretch out to a high degree, it will spring back as the tightened strings are allowed to renew their crazed contortions. The result is a gel. The presence of a few such strings in a fluid which otherwise contains only small water-like molecules is an astonishingly efficient way of turning a liquid into a solid. Think of a table jelly: we know that it consists mostly of water, because we put it there ourselves when dissolving the small blocks of coloured jelly cubes in water from the kettle. Yet after cooling in the fridge (but *not* the freezer—the water itself must still be liquid inside it) it remains resolutely solid, albeit with an occasional wobble. Although this is a normal experience and therefore no surprise, we really ought to be astonished at it. Turning liquid water solid by the addition of a small amount of sticky rubbery stuff starts to make us think about what really confers solidity to matter.

[3] But for readers without immediate access to a gentle stream of water and an electrostatically charged object, the water stream bends towards the object. This turns out to be a direct consequence of the electric nature of water molecules—they each have one side slightly more positive and the other more negative, so first orient in the electric field of the charged object and are then drawn *en masse* towards it.

What is really needed to make a solid is not space filled up with rigid stuff (though plenty of solids are like that—stone, metal and ice, for example), but rather space *connected* by sufficient paths that construct some sort of framework. This is after all how we make buildings solid, yet with plenty of room inside to move about in. The solidity is conferred by the planar walls, or in more modern buildings not even them, but the linear girders and struts that make up the steel scaffold of the building. Gels work after a more random version of the same principle, the giant string-like 'macromolecules' taking the place of girders, themselves taking up only a fraction of the space of the fluid, yet giving it a soft solidity.

So it was on an informed hunch that my colleague dropped in to talk about his surprising jelly—surprising because his student, a young Greek chemist, then working for her doctoral degree, had not added any stringy macromolecules to the liquid in her test tube at all, but only a small amount of 'peptides'. Now that word was my problem. Perhaps it is for you as well. I had studied very little biology, and although I had heard the word before, all it did was remind me of very complex biological ideas that I had never understood. Even scientists, you see, have problems with each other's jargon. To make science work, we have continually to admit our own ignorance to each other. It is one of our great levellers. Nothing moves on much if we do not pluck up courage to ask basic questions—exactly my mistake on that occasion. The result was that I understood nothing of the subsequent 'discussion' and for several weeks could not bring myself to go back to my chemistry colleagues to tell them that I would like to help if I could, and that I was very sorry, but would they please begin again in words of one syllable that a physicist would understand. But in the end the truth came out, and they patiently told me that a peptide was just a very short piece of nature's most common stringy molecules. These are the 'proteins' that appear everywhere in living systems. In normal usage we think about them as ingredients in foods of course—and for good reason. We build, heal and run our own bodies with protein molecules, so eating them is a good way of obtaining the molecular spare parts. But the peptides are much, much too small to have formed the sort of macromolecular jelly that would be needed to explain the rigidity of the obstinately non-fluid material in my colleagues' test tube. If we possessed eyes able to make out individual atoms, each peptide might look rather like the computer-constructed image of Figure 2.1

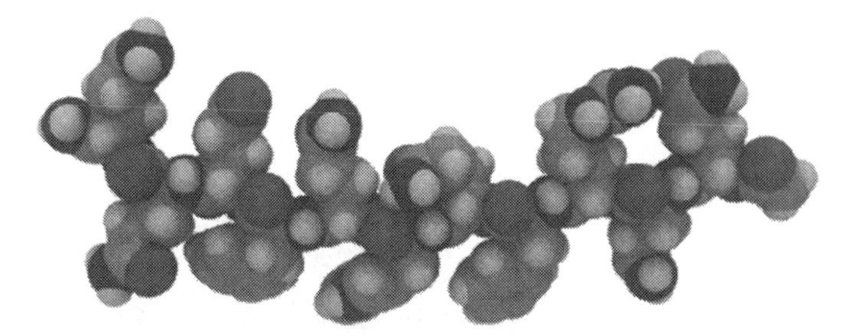

Figure 2.1 A computer-generated picture of one of the small peptide molecules. It is about 5 billionths of a metre long (courtesy of Dr. Sarah Harris, University of Leeds).

Once we had learned to think of them as tiny sticks we felt somewhat happier, if no nearer the answer to the puzzle. Only one explanation occurred to us: somehow the short peptide molecules must have joined up into much longer strings, themselves then getting tangled up at the junction points of a jellified network. Thinking of the peptides as just very short sticks was the first step, knowing that they must somehow have formed long strings was the second. The problem was that there was nothing any of us knew that would effectively 'glue' the peptide sticks end-to-end, back into the long protein strings that they once came from. There was one clue though: my chemist friends did know that peptides are somewhat 'tacky', not at their ends but along their sides. Hydrogen atoms, the very lightest sort, protrude from regular places along the length of the sticks, and can be attracted to oxygen atoms on neighbouring sticks. Might this be the cause of the formation of long string-like objects? Perhaps we should begin to call them 'tape-like' now, because joining hundreds of molecular sticks side by side makes an object that is indeed long and sinuous, but looking more like a minuscule model fence, or watch-strap, rather than a single rope. An imaginary computer construction of the sinewy tape, built out of the simpler peptide rods, would look something like the picture in Figure 2.2 Had the chemists discovered a jelly made of self-assembling molecular tapes? At last we had arrived at the key question that began to unlock the puzzle.

I was able to be of some help at this point owing to a very simple sum that I knew how to do. We had already used the existence of a jelly-like state that we could observe with our eyes to infer the existence of something we could not see: the molecular strings inside the fluid. Suppose we went on to measure the stiffness of the jelly—would that *number* tell

Figure 2.2 A computer-generated picture of a short section of the self-assembled twisted tape made of many peptide molecules (each one lies across the width of the tape). (courtesy of Dr. Sarah Harris, University of Leeds)

us anything more? If you have experimented with ordinary table jelly you will know that the more water you add, the more flexible and less stiff the jelly becomes. Once you know the conversion rate from 'amount of added string-maker' into 'stiffness of the gel', it turns out to be a simple matter to turn a measurement of the flexibility of the gel into one of the total length of string in the jelly. Now that is very helpful, because if the peptide sticks had joined end to end, they would have been able to make a much greater total length of string than they could of tape, by joining side by side. The difference was so big that it completely swamped any small errors of measurement, and the answer was very clear: the peptides had indeed spontaneously assembled side-by-side into tapes. Although we could not see them, we were sure that this rather novel form of matter now existed in our little glass containers. There was another clue: each tiny section of such molecular tapes would look very like a type of structure that many ordinary protein molecules make when they fold up upon themselves to assume their functional forms in living things. These tiny structures, bits of our extended fences resembling rickety wicket gates, are known as 'beta-sheets'—and leave a characteristic signal on infrared light passed through a solution of the proteins containing them. This uses only standard equipment found routinely in most chemistry laboratories. So it was a simple matter to make the infra-red light experiment and, sure enough, the beta-sheet signal from the gels was very strong.

Now it was time to write all this down and go public. It is an exciting moment for a research student: until that point research papers are difficult and impressive documents written by 'real scientists', distant, famous or intimidating, and published by volume in journals with impressive-sounding names. 'Could I ever persuade a research journal to publish something written by *me?*' is the question on a new student's mind. Close behind that thought runs the knowledge that a future career in science depends on succeeding at this exercise not just once but several times before the funding for the PhD research runs out. It is an exciting achievement to publish the next paper in a series of incremental advances, but much more to announce something completely new, just the offering we thought we had. Imagine, then, our dismay when the journal returned the work saying that it was inconclusive, lacked sufficient evidence and made unsupported claims. It was a hard enough blow for the two of us with permanent university positions, but much harder for a PhD student, who had no such security, only hopes that seemed rapidly to be dwindling.

Although we had by then faith in our theory of the self-assembled microscopic tapes, that faith had been based on an unusual combination of evidence from different sciences: chemistry, biology and physics in this case. Perhaps the problem was that any other reader without the experience of that combination would see not its strength but only the relative weakness of one of the strands of our story on its own? We needed more help to build a case for our idea—and before too long! Was there any way in which we could hope to 'see' the tapes in any direct sense? By enlisting expert help from other colleagues in the university, we were eventually able to find images, such as the one in Figure 2.3, of the long snake-like paths of the tapes using 'electron microscopy'. In this technique the carriers of electric currents inside wires, the tiny particles called electrons, are freed from their metallic confines and focused onto the object to be investigated. A microscope using electrons is able to capture much smaller details in an image than one using visible light ever could. The pictures improved, but like every new experiment brought more puzzles with them, even as they answered some of the questions. Sometimes much thicker and more rigid microscopic objects showed up than the flexible tapes we had seen at first. What could they be?

So well before one chapter of the exploration had been closed, another had opened. As things turned out, the microscopy was enough to

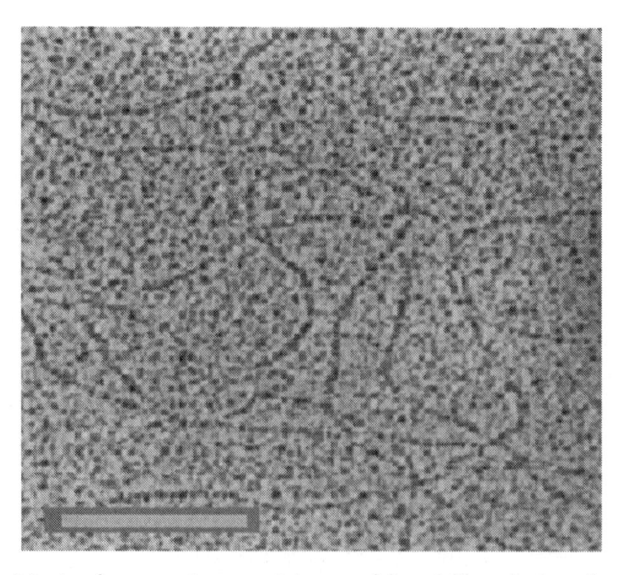

Figure 2.3 An electron microscope's image of the gel-like solution of peptides, after drying. String-like structures are clearly visible. Image courtesy of A. Aggeli, University of Leeds.

allow us to publish the first findings,[4] and the existence of the 'beta-sheet tapes' was soon confirmed by others in many different experiments. They showed up not only in solutions of the little peptide molecules we had used, but also in many cases of ordinary proteins at higher concentrations than would be found in living cells. These observations opened up a connection with medicine: some degenerative diseases such as Alzheimer's are not directly caused by bacteria or viruses but arise from a wrong turning taken in the assembly of proteins. Instead of folding neatly into their intended functional units, some proteins aggregate together in long fibril-like bundles that become denser and denser as the disease proceeds. The fibrils look remarkably like the more rigid objects that we had seen in some of our electron micrographs—and the connection is not accidental. Understanding, and if possible limiting the growth of these structures when they are damaging human tissue, is one of the exciting applications of the discovery of the tapes that is still active today.

Many other very different characters have contributed to the story: a married couple of Russian theoretical physicists spent some years with us while things were very hard in Moscow. They developed a mathematical theory for the self-assembly, including a prediction of the thicker fibrils and fibres. This allowed us to test the ideas again and again against measured numbers, and suggested new experiments all the time. Another young arrival brought a vibrant team of researchers skilled at using powerful computers to simulate complex molecules such as the peptide tapes, right down to the individual atoms. Among many young scientists who joined the growing team, one bright and cheerful PhD student explored the way that the tape-like polymers moved, and how that could control the way that the fluids containing them flowed. Tragically he suffered from a chronic disorder of the immune system and died soon after completing his PhD work. He knew all the time that he was living on a knife-edge but kept everyone challenged by his courage and his science. Perhaps it was his death that made me realise that this team, working together on winning a little more 'wisdom of natural things', had also been creating a community, trusting and appreciating each other's contribution,

[4] A. Aggeli, M. Bell, N. Boden, J.N. Keen, P.F. Knowles, T.C.B. McLeish, M. Pitkeathly and S.E. Radford, Responsive gels formed by the spontaneous self-assembly of peptides into polymeric β-sheet tapes, *Nature* 1997, **386**, 259–62.

supporting each other through difficulties, and sharing a common belief in the value of our common goal.[5]

The Mystery of the Ever-Present Motion

For our next story of science we travel back nearly two centuries to find a young Scotsman from Montrose with a passion for plants, on a ship bound for Australia. Robert Brown had managed to secure a berth in 1801 as ship's naturalist on the expedition of the *Investigator* to map the coastline of the still mysterious island continent. The mixture of relief, joy and excitement must have been powerful for the quietly spoken 24-year-old. If a successful PhD thesis and a collection of publications are the entry to a scientific research career today, then the chance to discover and catalogue new species in uncharted lands was the equivalent passport for an aspiring naturalist in the early nineteenth century. It would mean a 5-year absence from home, but the trip proved highly successful, for Brown possessed one of the most important scientific skills of all: a powerful and concentrated skill for observation. He did indeed assemble a wonderful collection of plants in Australia, and catalogued over 1000 new species, though must have suffered terrible disappointment when a large fraction of his carefully collected specimens that he had mailed home was lost in a shipwreck. In spite of this, once back in Britain, his fieldwork led him to a salaried position looking after a botanical collection in London. He was able to use this platform for 30 years to look hard at the detailed structure of his beloved plant species. He became one of the first naturalists to notice that plant cells seemed to possess a central particle, for which he gave us the term *nucleus*.[6]

It is the nucleus of all living cells that stores the biochemical instruction set of DNA, yet this highly significant breakthrough is strangely not the discovery for which Brown is best known—for that does not belong to botany at all, but to physics. Among his notes from 1827 were mentions of other particles in pollen grains that he had also studied intensely with his simple but powerful microscope.

[5] For a scientific summary of the research, see R.P.W. Davies, A. Aggeli, A.J. Beevers, N. Boden, L.M. Carrick, C.W.G. Fishwick, T.C.B. McLeish, I. Nyrkova and A.N. Semenov, Self-assembling β-sheet tape forming peptides, *Supramolecular Chemistry* 2006, **18**, 435–43.

[6] For further information on Brown, see D.J. Mabberley, *Jupiter Botanicus: Robert Brown of the British Museum*. London: British Museum (Natural History), 1985.

Intent as he was to understand the mechanisms underlying the role of pollen in plant reproduction, he was particularly taken with the constant jittery motion that the smallest particles within the grains seemed to possess. No matter how long he waited, the motion never seemed to die down. Was this vibrancy a clue to the secret of living matter itself?

He determined to trace the origin of the 'ever present motion' as he termed it. He had not, in fact, been the first to notice it, but was the first to let the observation turn him aside from his intended projects for a while. A willingness to follow the suggestions of circumstance that we have already seen can be an essential step in the discovery of new properties of matter. He became rapidly convinced that the eternal jiggling was a property of all particulate matter, independent of its substance, living or not. A patient and thorough series of experiments included water suspensions of particles from charcoal, chalk and rock dust, as well as plant material. Beautifully designed tests of various possible causes of the motion ruled them out one by one. Currents in the fluid excited by evaporation could not be the reason, since the motion persisted when the fluid was sealed. Electrical effects were equally improbable since he observed no change whether the apparatus was charged or not. Magnetic forces, vibrations of the room and the effect of light were, likewise, all systematically excluded. Readers of the monograph in which Brown published his findings in 1828 can feel themselves sympathising with his sense of dissatisfaction, even pain, that he could not produce a cause for the extraordinary microscopic motion. Today it is still difficult to publish 'pure' experimental results—a scientist is always expected to offer an explanation, even a tentative one. Brown clearly imposed that expectation on himself, as he sought again and again for the opportunity to suggest an underlying cause that was not immediately excluded by one or other of his experiments. All the more tantalising must have been his realisation that the motion is 'universal'—not depending on the particularities of particle or fluid.

Sometimes even the deepest questions simply arise before the time to answer them has come. One of the most impressive demonstrations of self-restraint within any scientific writing must be Brown's masterly scientific detective work, its long list of dead ends and his explanation of why he was not proposing a theory for the effect. In the supportive words of the great scientist and science communicator, Michael Faraday, which carry deep wisdom for all engaged in science:

Not asserting that a new power of matter was concerned; not denying that the powers with which we are acquainted might not be sufficient to originate the motion; but thinking it much more philosophical to acknowledge ignorance as to the mode of action in these cases, and to suspend the judgement, than, by the assumption of an opinion, which must have been hypothetical, run the great risk of shackling the mind by the admission of error for truth.[7]

Brown wisely guessed that satisfying the temptation to suggest various untested causes might well have set others along false trails before they had allowed imagination sufficient free reign. It is a wise application of insight into the way that scientists often work—single-mindedly pursuing an idea, when he suspected that the best thing to do in this case was simply to wait. Had he made any further suggestions, however non-committal, he might well have impeded progress towards finding the true cause.

One very far-fetched idea was in circulation. Faraday was much taken with Brown's work, and also distressed at the way that some of the journals of the day had misrepresented the 'Brownian motion' as indicating the presence of a 'vital force' within living matter (recall that this is one of the very first ideas that Brown ruled out). People then, as perhaps now, sometimes asked too much, too quickly, of science; it is never easy to live with open questions and always tempting to grasp for the support of easy answers. Faraday devoted the entirety of one of his popular and highly theatrical Friday evening soirées at the Royal Institution to stressing what Brown had *not* claimed, while urging the significance of his research. In his lecture notes for that evening,[8] there are only one or two statements written out in full that he did not want to forget or cut short. One was his castigation of the misrepresenting press—the other was a personal suspicion that the random motions might one day be able to shed some light on the 'molecular theory of matter'.

It might come as a surprise that as recently in scientific history as 1830 the existence of atoms and molecules was by no means universally accepted. True—they may be a convenient way of cataloguing the amounts of different elements that would react with each other chemically. Indeed the tradition of postulating their real existence may have originated with the ancient Greeks in the fifth century bc, but since no experiment could 'see' them directly it was hard to defend

[7] Michael Faraday, *Proceedings of the Royal Institution* 1829, **April–June**, 364.
[8] Michael Faraday. Royal Institution of London, ms. F4C, pp. 253–5.

their existence at the same level of meaning as, say, the nuclei of plant cells. Both Brown and Faraday were wise not to attempt a hastily conceived theory. Neither lived to see an explanation of Brownian motion, for it was not until 1905 that its connection with the existence of molecules was worked out mathematically by Albert Einstein.

Why the answer had to wait was determined more by the need for the right questions than the right methods of solution. They were among the questions that converged within Einstein's mind throughout that wonder-year.[9] The result of his extraordinary thinking in 1900–5 took the form of three revolutionary papers, through each of which he transformed our ideas about the world. It has sometimes been a challenge to see what, if anything, could have linked his foundational work on the structure of space and time, the foundations of quantum mechanics and the puzzle of Brownian motion itself, but one connecting motivation was a profound dissatisfaction with 'patchwork' theories of nature.

Einstein felt discomfort with the idea of one law to govern one aspect of the world, while a different law held elsewhere. For example, if there were laws of physics that applied to atoms, then they should apply to larger objects as well. He knew already about the second aspect of Feynman's miniature description of the atomic theory—that if these particles existed then they must be in constant yet random motion. He knew as well that this motion would generate the manifestation of the property we call 'heat' in collections of very large numbers of atoms. Now suppose a particle as large (compared with atoms that is) as one of Brown's pollen granules were placed among a collection of molecules in seething thermal motion. It, too, would have to pick up the random packets of energy that were continually jumping from particle to particle. During some short intervals of time a few more molecules colliding with the larger grain on the left than on the right would give it a tiny net jump rightwards. A little later the imbalance was just as likely to occur in the other direction. A few lines of algebra was all it took to show that the effect would be just the 'ever present motion' so meticulously observed in every ground-up mote of dust that Brown had studied. It is almost as if Einstein had invented a 'thought-microscope': motion that is visible, combined with a deep insight into how nature works, revealed the behaviour of structures too small to see.

[9] Abraham Pais, *Subtle is the Lord: The Science and the Life of Albert Einstein*. Oxford: Oxford University Press, 1982.

This story continues to the present day. In fact we have already met one of its later chapters in the account of the 'self-assembling peptides'. You might well have wondered how it was that any inert and senseless piece of matter could 'self-assemble' with any other. How could they even move towards each other without some guiding force? The answer is that they do it by Brownian motion. Random, blind searching by one small molecule for another will, if they are present in sufficient numbers, find partners. If at that point there are the right attractions present, then the partners will lock together and continue searching for others, until much larger structures emerge, such as the tapes and fibres. Life itself relies on Brownian motion to drive the processes within all living cells that keep them functioning as parts of muscle, skin, root or leaf. Or, for that matter, pollen—perhaps Robert Brown's willingness to leave his beloved botany for a while has, in the end, brought the treasures he unlocked back to our understanding of the living matter under his microscope.

Beneath the Surface of Solidity: the Bishop of Lincoln and Light

If there is one theme that unites the stories of science we have told so far, it is the desire of those who 'seek wisdom of natural things' to peer beneath the surface of matter, not only with their eyes, for our senses do not perceive everything, but in their imaginations and minds. To unlock an understanding of how nature works by the atomic theory, the twisting molecular tapes, the dance of the particles in the pollen—this requires the imagination of structures not seen, but whose presence can be conceived then connected to consequent phenomena on a scale that can be seen or measured. In each case the interplay of imagination, form and number play a part in sharpening or restraining the imagination towards the truth. Behind each story lies the urge to recreate nature within human minds, at face value a hugely ambitious project and one whose success could not possibly be guaranteed. Why should our practical human minds compass the ability to attempt such an extraordinary thing? To embark on it requires one of the actions of 'faith in science' we identified in the last chapter—faith that the attempt is worth making, that the hidden structure of the universe is not necessarily forever beyond our grasp. But it calls on another act of faith, not in our own abilities but in the type of the structure of the universe itself. We need to believe in a sort of *uniformity*, not in the sense that 'all things are the same' but that matter behaves consistently, not capriciously. As Isaac Newton put it, science works because 'a stone falls in Europe just as it does in America'.

It was Newton of course who epitomised the great flowering of mathematical physics in the Early Modern period. The seventeenth century saw the powerful uniting of all the threads that make up science, encouraged anew by a reinvigorated social standing for scientists, such as the formation of national scientific academies, and far-reaching new experimental instruments; among them the telescope and microscope. We could tell many stories from that period of course, of Galileo's revolutionary telescopic observations of the heavens and his complex relations with the church, of Robert Hooke and his beautiful discoveries in optics and microscopic life, of Descartes and his mathematical descriptions of space. But these are better told elsewhere, and for our purposes there is a danger in dwelling in a period of such fertility: for some have mistaken this period as the 'beginning of science'. Their accounts sometimes contain wordy and pointless discussions of who might qualify as 'the first real scientist'. From what we have already seen, though, we must treat such neat versions of the history of science with suspicion. For 'natural philosophy' seems to be an aspect of human nature that we are beginning to suspect runs deeper than the predominant outward form of any one age. So instead we continue our backward journey through time into the high Middle Ages of the thirteenth century. At one time thought of as an intellectual backwater of history, when the darkness of mysticism, magic and astrology spent centuries stifling the emergence of true scientific enquiry, it is now increasingly seen as the nursery of Renaissance thought, a bridge from the creative thinking of the ancients to science in its modern form.

Vitally for us, this is also a time when knowledge was far less artificially compartmentalised than now, and when learning was becoming open to new questions of nature and geography, as well as to the tradition of the ancient Greek and Roman authors. Also underappreciated in the West, the Middle Ages saw a particularly rich period of scientific understanding in the Islamic countries of Arabia, North Africa and Spain, to the extent that many of the best Latin translations of ancient Greek works available in thirteenth-century Europe had actually arrived at the northern European schools via earlier Arabic. Centres of learning were fewer than today, but by no means an insignificant component to international medieval culture. Our oldest universities were at that time still in their infancy, but other schools existed in the orbits of great cathedrals such as Chartres in France and Hereford in England, as well as in the monasteries.

It was into this world that the Englishman we know as Robert Grosseteste was born towards the end of the twelfth century. Other than a Norman provenance, his family origins are unknown, suggesting that his education was a charitable one and his background humble. However he came by his learning, we know that after early clerical positions, and possibly some diplomatic employment in Paris, he became Master to the Oxford Franciscans in 1228, then Bishop of Lincoln in 1235. He read, thought and wrote prodigiously about theology, science and philosophy, although it is important for us to realise that the boundaries between these disciplines that seem so clear to us were much more blurred then. The categories of questions and methods for answering them were still undifferentiated. This looks bewildering to us until we think for a moment about how much work it must have taken to find out where the natural boundaries lay. Grosseteste was one of the great contributors to this crystallisation of disciplines, but he took to his work a characteristic that spanned all of them—a desire for strong foundations in method and materials of thought.

His pioneering work on biblical texts is a good example: he bemoaned the poor quality of the Latin translations of New Testament biblical books on which most commentators based their analysis. Determined to recover the original sense as far as possible, he took pains to learn the Greek in which they were written—discovering along the way just how corrupted the multiple retranslations had become. How could one hope to preach anything but nonsense when empty words were piled up on the shaky foundations of mistranslations?

His passion for understanding the workings of the natural world around him was similarly coloured by a suspicion of rhetorical methods of argument that lacked foundation in observation. Textbooks that lean towards a neat and tidy interpretation of history will often record that 'the experimental method' was invented much later, in the Renaissance, but it is clear from reading perceptive thinkers like Grosseteste that he knew that truth is approached only when ideas and observations interact. Without requiring him to have formulated a way of doing science that would have elicited nods from the founders of the Royal Society over three centuries later, he nonetheless is able to write of how the causes of a spreading infection can be determined by testing each possibility practically and observing the outcome of the tests. As well as the cycles of living beings, he wrote about the stars, the motion of the Earth, colour and sound.[10]

The strong desire for what we might call 'foundational thinking' produced his most startling insights into the subject that today goes by the name of the 'stability of matter'. In his work on light, matter and cosmos, *De luce*,[11] Grosseteste demonstrates the sort of second reflection symptomatic of good science. He refuses to recognise what we experience as self-evident truth, but instead realises that questions lurk everywhere. Recall one vital property of Feynman's atoms—their mutual repulsion when close together. If this were not the case then solid or liquid matter as we know it could not exist—all matter would simply pass through itself. We are so used to the solidity of 'stuff' that for most of us its explanation is not a great puzzle. Yet think of a simple version of an atomic picture in which the atoms are really point-like particles—since this is the picture faithful to the ancient Greek *atoms*—the indivisible ones. At first it seems as if this might be a promising route to explain the matter we experience in terms of its hidden, and simpler, substructure—just the 'peering beneath the surface' that our stories of science have at their core. But in this case classical atomism does not work. If we stay with the idea of point-like particles then solidity simply does not appear.

Beautiful examples of just this sort of 'point-like matter' actually exist on a vaster size scale: for entire galaxies of stars like our own Milky Way are effectively massive bodies (the galaxies) composed of point-like particles (the stars). Stars are essentially point-like objects within an entire galaxy because the typical distances between them are so vast when compared with their sizes. Powerful telescopes have actually been able to identify a few cases of galaxies colliding (a beautiful example can be seen in Figure 2.4). Although their mutual gravity twists and wreathes giant streamers of millions of stars from

[10] For an introduction to Robert Grosseteste's science, see: A.C. Crombie, *Robert Grosseteste and the Origins of Experimental Science*. Oxford: Oxford University Press, 1953. For an example of recent literary and scientific analysis of a single thirteenth-century scientific work, see Greti Dinkova-Brun et al., *Dimensions of Colour: Robert Grosseteste's De Colore*. Durham: Institute of Medieval and Renaissance Studies, 2013.

[11] Cecilia Panti, Robert Grosseteste's *De luce*: a critical edition. In *Robert Grosseteste and his Intellectual Milieu: New Editions and Studies* (eds John Flood, James R. Ginther and Joseph W. Goering). Toronto: Pontifical Institute of Medieval Studies, 2013. An English translation of *De luce* by Neil Lewis accompanies the edition.

Figure 2.4 A pair of colliding galaxies imaged by the Hubble Space Telescope. They have already passed through each other. Image credit: NASA/William C. Keel (University of Alabama, Tuscaloosa).

their envelopes during the encounter, they really do pass right through each other without any of the stars themselves colliding at all.

So the particles that make up solid matter must additionally obey Feynman's repulsion rule, excluding others from a finite region of space. To explain bulkiness, the atoms cannot be just point-like but need to be bulky themselves. This leaves us with a problem: at this point have we really made any progress? We seem to be saying no more than to make solid matter we need atoms that are themselves just solid. What could the atoms themselves be made from that gives them this solidity? We seem to be caught in an endlessly recursive loop that never explains solidity by anything other than itself. This argument was actually one of the objections to the atomic theory of matter we saw in Robert Brown's story that kept it in doubt throughout the nineteenth century. The disputants had good reason: only the advent of quantum mechanics in the 1920s really solved the problem.

Robert Grosseteste was clearly well aware of this recursive impasse. The reason for his interest in the 'corporeity' of matter may well have been born from a study of the theological reflections on matter in the extraordinarily fertile thought of the twelfth century (but recall that 'theology', 'science' and 'philosophy' were not recognised as separate categories of thought at all in 1230). The structure of matter was important to medieval theologians for at least one vital reason: in the church's celebration of the Mass, the bread consecrated by the priest was believed to become, in some sense, the body of Christ—but in *what* sense? Matter taking on different apparent or 'accidental' forms was commonplace—water freezing or boiling, for example. Aristotle's physics, newly released into Europe in the twelfth century, contained the powerful idea that all matter had a 'substance' that was more permanent, if somehow more hidden, than its visible or tangible 'accident'. Such a theory of matter seemed to allow for the more unusual possibility that a substance might change (bread to Christ's body) *without* a change in accident (a loaf is still a loaf).

We have already seen how creative thought can arise when two powerful mental streams meet, such as Robert Brown's drive to understand the reproductive mechanism of pollen colliding with a tenacious pursuit of his strange observations of 'ever present motion'. In Grosseteste the colliding streams of thought are a theologically driven desire to understand how matter can be changeable yet solid, meeting a deep curiosity into the connection between matter and the utterly different thing we call light. Aristotle's framework might help to explain phenomena such as melting that we now call 'phase changes', but it did not address the problem of solidity of matter, nor the very opposite of solidity we experience in light's fleeting and weightless rays. Here is Grosseteste in his *De luce* (*On Light—or the Beginning of Forms*):

> *But a form that is itself simple and without dimension could not introduce dimension in every direction into matter, which is likewise simple and without dimension, except by multiplying itself and diffusing itself instantaneously in every direction and thus extending matter in its own diffusion.*

Here is the 'particle problem' once again—but this time with a very ingenious solution. Grosseteste observes that light, unlike atoms, *does* possess a natural 'extension'—open a shutter and it streams in to fill the dusty air beyond uniformly and immediately. If matter cannot of itself (simple and without dimension) fill space, then maybe the operation of light *on* matter might endow the tiny particles with extension by carrying them, or somehow breathing extendedness into them. Actually, he is very careful to say that this source of corporality might not actually be light itself, 'but if not then something very like it'.

This is breathtakingly deep stuff, not only because it is clear that Grosseteste actually observed the properties of light carefully (he described the angles of refracting beams through glass and drew attention to the power of mathematical geometry to describe light's properties). More significant is his early insight into the need for something much more than classical particles to explain the solidity of extended matter. We are so used to sitting on chairs and not falling through them that we do not perceive our lack of true understanding of such mundane experience. Realising that there is a problem requires hard thinking about what infinitesimal particles really will and will not explain. Remarkably, Grosseteste's insight turns out to be more or less correct. The reason that light is space-filling is that it is a wave—a sort of mathematical opposite of a particle. But matter also has a wave-like aspect, unremarked until the revolutionary discoveries of quantum mechanics in the 1920s. It is precisely the quantum waviness of matter that allows it to be solid, and prevents my falling though the chair I am sitting on like two galaxies falling through each other.

De luce then contains an astonishing logical move that would have delighted a great unifier like Einstein. Grosseteste is faced with one serious problem in Aristotelian science, one that he cannot accept as a Christian, even while he takes all of the pagan philosopher's methodology as his own. For Aristotle would brook no 'first cause', and in consequence his cosmos could have no beginning. Without a creator there is no moment of creation. In a remarkable symmetry to the proposition of the 'steady state cosmos' by Bondi, Gold and Hoyle in the twentieth century (like Aristotle these authors wished to avoid a universe with a beginning as it fitted too comfortably with theism), Grosseteste needed a physical account of the origin of the entire cosmos. To formulate it he makes an extraordinary leap of the imagination: he attempts to apply his theory of local matter to the structure of the universe as a whole. Beginning with a flash of light, the entire universe is filled and expanded by its self-propagation until it has reached huge dimensions. Solidifying in its exterior shell, re-radiated light from this shell of 'perfected matter' then concentrates matter back towards the centre, leaving the successive planetary shells in its wake, and the unrefined elements of fire, air, water and earth at the centre. Of course this 'medieval big bang' theory of the universe is a theory of the Aristotelian cosmos: the Earth is in the centre with the Sun, Moon, planets and stars circling around it. We know now of course that this is wrong, but it would take another three centuries, more remarkable minds and observations of very subtle planetary motions to begin to shift the massive intellectual weight of such an Earth-centred cosmology.

What is extraordinary in this thirteenth-century mind is the realisation that we can conceive the underlying structures of nature, from the smallest components of matter to the entire cosmos, and in a single sweep. But wisdom of natural things takes time—and a historical community of people—to acquire. In Robert Grosseteste we see the desire to peer beneath the surface of phenomena, harnessing both keenness of observation and power of thought to reach an insight that only came to full flower eight centuries later.

On the Nature of Things: Bede's Science

The thirteenth-century flowering of scientific thought in which Grosseteste embarked on his outstanding investigations did not appear from nowhere. The classical tradition of Plato had been preserved (or at least its core) throughout the first millennium in the Western monastic tradition. Second, as we saw, the twelfth century had seen a rediscovery (from the point of view of the Roman West) of the works of Aristotle, thanks to the Islamic civilisation of North Africa and Spain, which had preserved and commentated on his natural philosophy and logic in a fruitful scientific tradition of four centuries from AD 800. A third strand of influence seems to have been constituted in a tradition of scientific writing with the generic title *De Natura Rerum*—On the Nature of Things. These Latin texts from the western part of the Roman Empire (Lucretius famously wrote the first example) and later the Europe of the early Middle Ages constitute a continuous if tenuous bridge between the ancient world and the renaissance of wider learning in the later Middle Ages.

Perhaps the most fascinating example, and one that takes us a further leap back in time another five centuries, is due to the celebrated northern English scholar, the Venerable Bede. His tomb, visited by many people today, is found in the stunning Norman Galilee Chapel of Durham Cathedral, a few miles from the site of the monastery at Jarrow where he spent most of his life. Best known as the first post-Roman British historian (his *Ecclesiastical History of the English Speaking Peoples* has been translated and edited many times), he is less familiar for his writing on science and mathematics. Indeed, it has taken over 13 centuries for an English translation of his major scientific works to appear.[12] Written in

[12] The first English translation and commentary on his *De Natura Rerum* and *De Temporibus* can be found in Calvin Kendall and Faith Wallis, *Bede: On the Nature of Things and On Times*. Liverpool: Liverpool University Press, 2010.

around AD 700, the two short documents *De Natura Rerum* (On the Nature of Things) and *De Temporibus* (On Times) give the lie to any misconception that the darkness of the Dark Ages refers to anything but our ignorance of them. Written as introductory treatises for students at Jarrow, they also tell us how central to the monastic curriculum was learning about nature at that time. Nor do they represent a fixed tradition of preservation, but show evident signs of development.

De Natura lays before the reader an encyclopaedic coverage of natural phenomena, starting with the heavens, the stars and planets, then descending through the phenomena of the atmosphere to the physical geography of the Earth, its rivers, seas and mountains. It is a delight to read into an early seventh-century mind at work—trying to systematise a world of phenomena, many of which (like volcanoes) he has only heard or read about. A delicious passage contains Bede's discussion of why the sea remains salty, even though fresh water continually flows into it from all the world's rivers. This is an old problem which had given rise to two hotly contested solutions. Pliny the Elder, whose *Naturalis Historia* (AD 77) is extensively quoted in Bede's *De Natura*, favoured the hypothesis that the fresh water failed to mix with the salty sea, but simply returned to the mountain springs by way of underground culverts. Bede argues forcibly against this in favour of quite another process:

> *Because the sea is not increased by the inflow of rivers, they say that the fresh flow is naturally consumed by the salt waters, or that it is carried off by the winds or by the heat of the sun, as we prove in lakes and ponds that are dried up in a brief period of time, or even that it reflows by a hidden passage into their own springs and runs back by the usual way through their own streams. But fresh waters flow above sea waters, because they are lighter; the latter certainly, being of a heavier nature, better sustain the waters poured over them.*[13]

A much more likely route to the mountains is by evaporation and precipitation, says Bede, and this is because of his experiential knowledge that fresh water will not sink through the denser saline of the sea. Bede is fearless in taking on an ancient authority in the light of observation and reflective thought. Any scientist would recognise his methods and motivation. This is not to say that he is a scientist in the developed modern sense, any more than was Grosseteste. But it does affirm that he

[13] Calvin Kendall and Faith Wallis, *Bede: On the Nature of Things and On Times*. Liverpool: Liverpool University Press, 2010.

values digging below the surface of phenomena, has some idea of what might constitute 'explanation' of natural phenomena, and furthermore sees this entire process as central to learning, as a natural part of being human and of his own vocation.

Bede's *De Natura Rerum* makes another move that is very significant, but only visible in the context of the works from which he draws. His is by no means the first Christian example of the genre of Lucretius and Pliny. He quotes more from Isidore of Seville's *De Rerum*, from the previous century, than he does directly from Pliny. But, as Faith Wallis has pointed out in her and Calvin Kendall's delightful new edition and commentary,[14] he strips out the extensive moral and spiritual allegorising that Isidore projects onto his encyclopaedia of nature. For Bede, his task as a Christian teacher and pastor is to seek explanation of nature on its own terms. He sees this achieving two things—it refutes superstition (of a natural world full of spirits and dark menace) and engenders an appreciation of wonder. It might also resonate with deeper theological motives. For a hint of those we need to travel back further to the complex world of clashing cultures that was the Eastern Roman Empire of the fourth century.

At Macrina's Bedside: Gregory of Nyssa and the Healing Reality of Mind

For our final story of science we travel earlier still, by nearly another three centuries, and to one of the most extraordinary documented conversations of the ancient world. We arrive in another period and place rich with debate, argument and intrigue: the world of the Christian Mediterranean and Near East of the fourth century. After more than two centuries of development under both persecution and official sanction, the appearance of alternative interpretations of the lively young faith was requiring a definition of orthodoxy. The explosive events behind the formation of the early church itself had been powerful enough for it to take that long for the doctrinal dust to settle. Great 'councils' of bishops convened from the known world to hammer out ideas such as the Trinity (how can one God exist in the three persons, Father, Son and Spirit?) and the relationship between the Old and New Testaments. Great intellects were required in

[14] Calvin Kendall and Faith Wallis, *Bede: On the Nature of Things and On Times*. Liverpool: Liverpool University Press, 2010.

the stretching of both thought and language to breaking point, driven by the need to describe the heady mix of new beliefs, new codes of behaviour and new ways of constituting communities that the last 10 generations had experienced. Consummate mastery of both language and diplomacy was demanded in the drafting of the early 'Creeds', or statements of belief. One such is the famous Nicene Creed, formulated at a council in Nicea in 325 CE.

Of the three theologians, collectively known as the 'Cappadocian Fathers', who were at the centre of these events, two were actually brothers—Basil the Great and Gregory of Nyssa. Gregory, it seems, was rather a reluctant bishop, but accepted the post at Nyssa at a time when his elder brother was hard pressed and needed authoritative support close at hand. Less well known is a figure deeply revered by both Basil and Gregory: their elder sister, Macrina. So well read and intellectually gifted was she that they referred to her as the 'Teacher'.[15] Historians of the church have remarked that one of the real Cappadocian Fathers was actually a 'Mother'! Gregory in particular enjoyed intellectual sparring matches with his sister. One of these arguments is recorded by him in loving detail—perhaps because it was to be their last: it took place at Macrina's bedside as she lay dying. Gregory records that, troubled by the sight of his distress at her suffering, Macrina suggests they leave grief to one side for the moment and occupy their minds with the sort of theological debate they used to relish. The result is Gregory's short book *On the Soul and the Resurrection.*[16] It is a discussion with a very modern ring to it—essentially Macrina suggests they debate the reality of the human mind (although usually translated 'soul', it is clear from the context that 'mind' brings a more faithful set of ideas to her meaning within today's language world). Are our thoughts really those of an independent agent, or just the epiphenomena of a turbulent material brain? She suggests that Gregory argues against the reality of mind, while she defends—and then they go to it.

At the end of a long list of supporting arguments comes a surprise: her crowning evidence that minds really exist draws on what we see when we hold an upturned bottle underwater. The passage is worth setting down in full:

[15] See, for example, A.M. Silvas, *Macrina the Younger. Philosopher of God*. Turnhout: Brepols, 2008.

[16] Gregory of Nyssa, *On the Soul and the Resurrection*. Transl. C.P. Roth. New York: St. Vladimir's Seminary Press, 1993.

> *It is by an abuse of language that a jar is said to be 'empty'; for when it is empty of any liquid it is none the less, even in this state, full, in the eyes of the experienced. A proof of this is that a jar when put into a pool of water is not immediately filled, but at first floats on the surface, because the air it contains helps to buoy up its rounded sides; till at last the hand of the drawer of the water forces it down to the bottom, and, when there, it takes in water by its neck; during which process it is shown not to have been empty even before the water came; for there is the spectacle of a sort of combat going on in the neck between the two elements, the water being forced by its weight into the interior, and therefore streaming in; the imprisoned air on the other hand being straitened for room by the gush of the water along the neck, and so rushing in the contrary direction; thus the water is checked by the strong current of air, and gurgles and bubbles against it.*[17]

We recognise straight away what is going on: this is science in the raw! Deep observation and contemplation of a visible phenomenon in nature leads to a deduction about the structure of something *unseen*—in this case the existence and elasticity of air. Had the bottle really been filled with nothing, the water would have rushed smoothly in to every corner. Instead Macrina draws our attention to the complex fluid dynamics generated by the clashing countercurrents of water and air, which can only change places by competing for passage through the narrow neck. It is an example as clear as that of any of the others we have examined, of 'looking' with our minds—Macrina calls it 'the eye of the experienced'—beneath the surface of reality to discover and become familiar with an aspect of its hidden nature. That is why she employs it as her *coup de grâce* in this final debate with her beloved brother: only *minds* may peer beneath the surface of matter and perceive there what is unseen by the eye alone.

She refers to other examples, notably the changing phases of the Moon. Rather than assuming that the Moon waxes from crescent to full because some substance is mysteriously added to it, we understand it to be a sphere illuminated from a changing angle by the sun. But since this is not something that we 'see' directly, the realisation must come from elsewhere. Like the bubbling air, the spherical Moon is another reconstruction of external nature within our own minds.

Here in as sharp a focus as we could wish for is a thoughtful and beautifully explained series of scientific discussions in a document

[17] Gregory of Nyssa, *On the Soul and the Resurrection*. Transl. C.P. Roth. New York: St. Vladimir's Seminary Press, 1993.

from the embryonic years of Christian culture, fuelled by the highest scholarship of the day, but earthed in a movingly personal setting. Macrina and Gregory throw a new light on our search for meaning to 'faith and wisdom in science'. Recall that Macrina's argument is working on two levels: she wants to display the eternal character of our minds by showing that they sit above material phenomena, but she is also conducting the whole debate to help Gregory cope with their shared pain, and to prepare him for their parting. Her faith is that this contemplation of wisdom about the natural world will play a part in the healing of her brother's pain. The entire book is well worth reading, for the natural way that faith in science arises within a theological discourse. Macrina and Gregory embrace a Christian thought-world at complete ease with the material universe and its exploration. Natural philosophy is part of the theological story that they are not only telling, but living.

Unsurprisingly from its title, *On the Soul and the Resurrection* eventually arrives at the theme of hope: again at the two levels on which the argument is playing out. Macrina wants to remind Gregory of the reasons to hope—it is part of the healing of his pain—but just as the brief salve for their suffering was drawn from the well of our relationship with the present physical world, so she evokes hope from the deeper, yet thoroughly physical Judaeo-Christian hope of resurrection and renewal of the physical order.

We have now glimpsed five settings, across two millennia, within which we find individuals and communities searching for the 'wisdom of natural things'. Although no such limited selection could possibly pretend to represent the vast cultural history of science, it does I hope communicate a flavour of what *doing* science is like. Let us draw a few threads together.

First, doing science is very old. Thinking about what might lie beyond and behind the superficial appearance of the world in a way that allows a different kind of 'seeing' takes us at least as far back as the fourth century. We could have carried our journey into the past on for many more centuries, to pick up the same thread in ancient Greece. We already know that we would find there, among many other things, the roots of the atomic hypothesis itself. Because the appearance of today's science is so coloured by the technology that supports it, we too easily fail to recognise the fundamental signals that tell us when people are doing natural philosophy. We also commonly fall into the trap of forgetting the long history of imaginative science which constitutes the heritage on which we build.

Second, science is a deeply human activity. Our modern 'professionalisation' of science tends to narrow our definitions of it as effectively as our modern technology. Real science can take place at an elderly woman's bedside, in a medieval bishop's house and on a sailing ship anchored by Van Diemen's Land just as much as in a modern university laboratory. It can also engage extraordinarily different people, personality types and backgrounds, creating colourful communities as it does so.

Third, science is more about imaginative and creative questions than it is about method, logic or answers to those questions. Asking why water flows turbulently, why the sea is salty, why matter should be solid, what causes tiny particles to diffuse rather than rest, what component of a fluid might have assembled into strings—none of these was an obvious question at the time it was asked, yet all provided the keys to a new deeper view of the natural world around us. They suggested ideas and experiments that led, sometimes sooner, sometimes much later, to the answers. Finding such key questions is a deeply imaginative act.

Fourth, science can be painful. A clearer insight into our world is not easily won. Hopes that we are on the brink of answers are often dashed. Even when we really are in possession of a new insight, persuading others that there might be a new way of thinking about peptide solutions, finding evidence that molecules are real, suggesting underlying links between light and matter, or thinking about minds and brains, is hard and painful work. We have more than once seen science accompanied closely by elation and disappointment, by life and by death.

Finally, the relationship between 'faith' in all of its connotations and 'science' is a long and rich one. Faith in newly formed and still awkward ideas is indispensable if we are to see further below the surface of our world. Faith in other members of the community of searchers after natural wisdom is equally vital, whether these are collaborators, readers or, as in Robert Brown's case, those hoped-for scientists of the future who would one day answer the questions that he was painfully unable to resolve. But we have to capitalise Faith as well, to do any justice to our historical and human analysis of its relation with science. The deeper we have probed into its roots, the clearer becomes the theological background to every aspect of its nature. For Gregory of Nyssa, science, soul and resurrection belonged within the same train of thought. Robert Grosseteste applied the same mindset to urging a deeper biblical theology as he did to a deeper understanding of solid matter. Robert Brown's honesty, and the huge respect in which he was held by the scientific community, was built on a foundation of Scottish Presbyterianism.

We are faced with an apparent head-on collision. In the face of demands from writers like Dennett and Dawkins that we recognise 'science' and 'theology' as incompatible activities, we begin to find this unlikely to have been the case throughout the vast majority of human history. By actually working through some real science ourselves, so that we are reminded what it 'feels like', we have found it to run rather deeper, and to touch more nerves, than these writers would have us notice.

But before we hasten towards any conclusion, we ought to do with theology what we have done with science—we need to try actually doing some to see what it 'feels like'. In the next chapter we will embark on a survey of biblical material that would certainly have been part of the world-view of Gregory, Bede, Grosseteste and Brown. Does this tradition say anything about how we 'see' or search into the physical world? If it does, is the pattern that emerges compatible or incompatible with the structure of science that we have drawn out in the present chapter? And might a search into the streams of narrative that once fed these great thinkers provide us with any resources to make sense out of the confused voices around science's place in society in our own time?

3

Creation, Curiosity and Pain: Natural Wisdom in the Old Testament

Search a concordance (a type of exhaustive word index) of the Bible for 'science' and clearly you will not come up with anything. But by now it is clear to us that this does not mean that the Bible writers have nothing to say about human exploration of the natural world. We have already identified this cultural project as very ancient—the concordance just confirms how recent is the language we currently use to talk about it. The same problem would arise were we to search for Biblical views on, say, 'gender'—there are no neat sections in the Old Testament histories, wisdom literature, prophets, gospels or letters of the New Testament under that contemporary heading, yet all of these different types of Biblical literature contain fascinating stories and discussions of men, women and the ethics of the power-play between them.

Here our language work will pay off richly. We know already that we should focus not so much on particular words but on material that, like Macrina's discussion of the water jar, reflects a mind seeking to establish links with George Steiner's 'inhuman otherness of matter'. Doing this saves us from another mistaken assumption suggested by some of the shrill arguments we encountered in Chapter 1, namely that biblical encounters with science begin and end with the creation story in the first chapter of the book called 'Genesis'. We will find instead that the Bible is shot through with different ways of talking about the origin of the material world, each using its own imagery. We will also find that the creation stories are only a starting point in the thread of 'wisdom to do with natural things', and that the Genesis accounts make up a very small fraction of the whole.

So at first sight this chapter seems to take us in a very different disciplinary direction, from accounts of scientific motivation and reasoning to a selective exploration of texts from a foundational religious tradition.

But let us suspend our prejudices and customary compartmentalisations for the moment, and continue the journey we began in the last chapter into the human engagement with nature.

Wisdom's Childhood: Proverbs 8

We begin with a text that is probably older than that of Genesis, and from a different stratum of literature—the 'wisdom' corpus. Assembled during the first millennium BC by royal and priestly writers and editors in ancient Israel, they reflect both the inherited philosophical traditions of Babylonia from the east and Egypt from the west. The wisdom texts also add a new distinctive voice: that of a monotheistic people whose God had endowed their relationship with him with a special historical significance. The extended hymn to wisdom itself, which we know by the title 'Proverbs', begins by advising the reader to listen and to acquire wisdom, for it is 'more precious than rubies'. But after a while a remarkable thing happens as wisdom *herself* takes on a female persona and a distinctive voice within the text, building to this beautiful crescendo in chapter 8:

> *The Lord brought me forth as the first of his works, before his deeds of old;*
> *I was appointed from eternity, from the beginning, before the world began.*
> *When there were no oceans, I was given birth,*
> *when there were no springs abounding with water,*
> *before the mountains were settled in place, before the hills, I was given birth,*
> *before he made the Earth or its fields, or any of the dust of the world.*
> *I was there when he set the heavens in place, when he marked out the horizon on the face of the deep,*
> *when he established the clouds above and fixed securely the fountains of the deep,*
> *when he gave the sea its boundary so that the waters would not overstep his command,*
> *and when he marked out the foundations of the Earth.*
> *Then I was the craftsman at his side.*
> *I was filled with delight day after day, rejoicing in his presence,*
> *rejoicing in his whole world and delighting in mankind.*

What a playful, delightful description of a young world full of hope! The storyteller seems to be one of two characters present at its creation. A creator ('the Lord') is certainly in the background, but centre stage is the fleeting, dancing character of Wisdom, who weaves around the components of the physical world, responding to their coming into being with joy. Although very old, this complex passage already contains several themes that reappear in, or lie closely below the surface of, a long skein of creation literature in the Bible. At its heart it embeds a sort of 'formula' for describing the world's creation: one that concentrates primarily on establishing boundaries: '. . . *marked out the horizon . . . established the clouds above . . . gave the sea its boundary . . . marked out the foundations of the Earth . . .*'. The vital aspect of creation in this tradition is not so much the naked existence of matter, but its *order*: the sky above, the sea separated from the land and the depths of the Earth below our feet. The 'taming' of the sea is especially important—a dangerous and alien medium is kept at bay almost as one would a tethered beast. Even today we know and fear the dangerous power of the storm, the shipwreck and the tsunami. Small wonder that in ancient cultures the ocean represented, by extension, forces of darkness and chaos that threatened to encroach once more on the land of the living. So emerging from this description of creation itself is a seamless account of our relation to it—no creation story is ever objective because we who tell the tale are part of the same ordering of the world, and survive only because of its stability.

In the same way, 'Wisdom' in Proverbs is no abstract philosophical category; rather it is an intensely practical method of living most fruitfully in the real world—one that avoids, as far as it is possible, its pitfalls and pains. So practical indeed that Wisdom is given a personal voice. It, or rather 'she', is deeply *relational*—'how do we respond to the created world?' is just as pressing a question for Wisdom as 'what is the structure of the created world?' Here, too, Proverbs surprises us, for the response of Wisdom is not principally one of gritty survival or the promotion of a moralising ethical framework, but one of childlike *delight* in the world. The Hebrew translated as 'rejoicing' can also indicate 'playing'—one is tempted to think of Wisdom at the dawn of the world as a little girl playing in its nascent streams and fountains, rather than as the grown woman who later in the book offers her guiding principles for adult life within established communities. So Wisdom herself is endowed with a narrative history. As with many important structural ideas in Judaeo-Christianity, she is not static and exterior to the larger biblical story of the relation between God and humankind. Instead, she is allowed right from her first entry to develop within it. As we will see, she does not stay a child for long.

The Creative Word: Psalm 33

Perhaps the most familiar Old Testament wisdom book is the collection of hymns, poems and prayers we call the Psalms. Many of them take the physical world as a springboard for praise, lament or pleading, and within these categories are several accounts of the world's creation. We do not normally think of the Psalms as a source of creation stories, but that may be because we are not alert to the variety of linguistic forms that differ from, and in most cases are simpler than, the Genesis versions. They are also frequently very short, and embedded within other material. This makes the creation stories of the psalmody easy to miss. But if we are alert to the structure of Wisdom tradition creation, as in the example from Proverbs, we will notice them frequently.

A good example is Psalm 33. It begins with a musical call to praise and finishes with an extended affirmation of hope for the nation chosen by God. But as soon as the 'word of the Lord' is introduced as a guiding principle for righteousness and justice (v.4), its effect is given a much grander scope (vv.6,7):

> *By the word of the Lord were the heavens made,*
> *their starry host by the breath of his mouth.*
> *He gathers the waters of the sea into jars;*
> *he puts the deep into storehouses.*

This is new vocabulary for the creative act. More specific than the 'establish', 'set', 'gave' or 'mark' of the Proverbs passage, Psalm 33 identifies the central creative act as that of speaking. The communicated word that sets the course of righteous jurisdiction for a nation becomes the word that orders and sets bounds to the chaotic forces (especially the 'waters of the sea') of nature in creation itself. Probably one should not seek to make much of the distinction between 'word' and 'breath' as applied separately to the heavens and the stars—this pattern of poetic parallelism is common in Hebrew literature, but the double usage does serve to bring two narrative lines of thought together within a very short space. For if 'word' brings the idea of principle or law, then 'breath' carries the notions of 'spirit' and 'wind' (all three are *ruach* in Hebrew). When we look later at the Genesis creation stories we will see how the two creative forces of word and breath have by then developed into a complex dual tradition. Here we see a more condensed, and very possibly earlier, version.

There are two other important roles played by this short creation story in Psalm 33. The first is evident only in the Greek translation of the Hebrew Old Testament known as the Septuagint, in use from the third century BC.[1] For there the translation used for 'word' was *logos*, used later very significantly, as we shall see, by St John in the opening passage of his gospel. By the time of the Septuagint, *logos* was already beginning to be loaded with Greek Stoic influence, carrying the significance of a universal ordering principle by which matter took form.

The second way that Psalm 33 plays out is to illustrate a rather general pattern in biblical references to nature in both Old and New Testaments. The passages never stand alone—they always seem to perform the work of a bridge between past and future, despair and hope, threat and deliverance, pain and healing. Here, as we have noted, the ordering of creation in the heavens and the waters translates the act of worship into an expectation of deliverance of God's people from distress. By way of comparison, in Proverbs 8 the same underlying narrative transports the reader from the follies of sloth, corruption and adultery to the attributes and actions of a life lived wisely. Thinking about nature is not, in the Bible, a 'spare-time' activity—it is part of the healthy life of individuals and communities, and part of the stories that carry them from their pasts towards their futures.

Dynamical Creation: Psalm 104

Not all the creation accounts in the Psalms are as condensed as that in Psalm 33. Many of the structural ideas we found in Proverbs 8 also reappear, for example, in the introductory verses of a longer narrative in Psalm 104:

> *Praise the Lord, O my soul.*
> *O Lord my God, you are very great; you are clothed with splendour and majesty.*
> *He wraps himself in light as with a garment; he stretches out the heavens like a tent*
> *and lays the beams of his upper chambers on their waters.*
> *He makes the clouds his chariot and rides on the wings of the wind.*

[1] The empire of Alexander the Great and his successors established the 'Koine' Greek of the time as the lingua franca throughout the Mediterranean and Middle East, which it remained throughout the early centuries AD.

He makes winds his messengers, flames of fire his servants.
He set the Earth on its foundations; it can never be moved.
You covered it with the deep as with a garment; the waters stood above the mountains.
But at your rebuke the waters fled, at the sound of your thunder they took to flight;
They flowed over the mountains, they went down into the valleys,
You set a boundary they cannot cross; never again will they cover the Earth.

Just as in Proverbs, we see the fundamental structures of the heavens, the Earth's foundations and the waters crafted and set deliberately in place. Again the writer stresses the importance of boundaries, and especially the boundaries between sea and land. The implied painful community memory of a former breaking of that boundary is only obliquely alluded to. It is not obvious whether this is referring to the story of the great flood in Genesis (and its many parallels in other literature of the first millennium BC in Babylonia and elsewhere), or merely to an imagined stage in creation itself before Earth's waters and land has been separated. But the same insistence that creation is primarily the establishing of order is as foundational in the Psalms' creation stories as in Proverbs.

Psalm 104 now seems to take over the story where Proverbs 8 had left off, working up the theme of a well-ordered creation by showing in detail how water, Earth and air can function creatively when they touch at boundaries, rather than overlap. Rains bring water to animals, birds and plants, but only when the dynamical cycle of water is controlled (vv.10,11):

He makes springs pour water into the ravines; it flows between the mountains;
They give water to all the beasts of the field; the wild donkeys quench their thirst.

The once static, homogeneous universe is no longer so dully uniform. By the same token it is no longer motionless either—new rhythms appear that connect the dynamics of the heavens to the cycle of life on Earth (v.19ff.):

The moon marks off the seasons, and the sun knows when to go down.
You bring darkness, it becomes night, and all the beasts of the forest prowl.
The lions roar for their prey and seek their food from God.
The sun rises, and they steal away; they return and lie down in their dens.

Then man goes out to his work, to his labour until evening.
How many are your works O Lord!
In wisdom you made them all; the Earth is full of your creatures.

We are presented with a remarkable picture of the harmony of nature, but a harmony that emerges intriguingly from the theme of *wisdom* we recognise from writings like Proverbs—the wisdom that marks a true understanding by humankind of its proper relation to the rest of physical creation. Psalm 104 sees the *work* that orders daily human activity within such a relational framework.[2] We might be forgiven for thinking this happy picture a little rose-tinted were our reading to stop there. But just as the writer of Proverbs knew full well that wisdom does not always rule humankind, so the Psalmist knows that the ordered balance of nature itself is a fragile one (v.29ff.):

When you hide your face, they [the animals] are terrified;
when you take away their breath, they die and return to the dust . . .
. . . he who looks at the Earth, and it trembles,
who touches the mountains, and they smoke.

Death, pain and chaos are not overcome, but are present to everyone and everything in the physical creation. The choice of phenomena such as volcanic eruptions and earthquakes of the inanimate world to partner death within the orders of living beings presents a stark structural contrast with the use of Moon and Sun in the portrayal of the rhythms of life. The monthly phases of the Moon are understandable, and actually predictable. Even the subtle pattern of lunar and solar eclipses had been known to civilisations much more ancient that that of Israel. But earthquake and volcano alike defy prediction to this day. They are emblems of the class of phenomena now formally, even mathematically, classed as 'chaotic'. Not only do people themselves depart from wisdom, with painful consequences, but creation also embodies disorder, departure from harmony and suffering for the creatures it cradles. There are no answers in the psalm—it is as if the question itself is only beginning to form, but if Proverbs' story of creation gave us a glimpse of the years of Wisdom's innocence, then in the Psalms creation returns with the equivocal lessons of experience in a painful world growing

[2] A similar parallel is set up in Psalm 19 between the ordered laws of the heavens and an earthly human community well ordered by wisdom, this time flowering within *law* rather than *labour*.

older. If the story of nature is intimately and irretrievably woven into the story of the people who live within it, then we can begin to see that physical and mental pain will inevitably be part of that relationship.

Creation and Correction: Jeremiah

The Prophetic books constitute a very different type of literature from the Wisdom writings of Proverbs and Psalms, but they also frequently feature creation stories. The prophets' task is often to criticise the nation of Israel when she chooses paths other than that of God's Wisdom, and their writings are deeply rooted in the historical events of their times. For example, events such as the destruction of the temple by Babylonian forces in 587 BCE and the consequent exile to Babylon of a large number of Judeans were interpreted as a judgement on the corruption of both temple and nation. The prophet Jeremiah's career spanned the period of Judah's history both before and after this exile—he seems to have written in both Judah and Babylon. The greatest issue that, for Jeremiah, had brought the exile onto God's people was the sin of *idolatry*: the worship of statues and other images made by human hands, representing the gods of surrounding nations.

It may be hard to understand today why this practice was either so very tempting or so very wrong in the eyes of the Old Testament prophets, when we tend to come across it today in the form of quaint local rituals featured in tourist guides of Asia. But from the 10 commandments of Moses, through the entire history and prophecy of Old Testament Israel and into the literature of the early church, idolatry is always presented as the principal obstacle to faithful engagement with the living, creator God. To entertain it was to remove the foundation on which covenant faith is built.

So it is perhaps not so surprising that Jeremiah presents another explicit *creation* story when confronting idolatry (in chapter 10, closely paralleled by a later passage in chapter 51), framed with an announcement of the old warning:

> *Do not learn the ways of the nations or be terrified by signs in the sky,*
> *though the nations are terrified by them.*
> *For the customs of the peoples are worthless; they cut a tree out of the forest,*
> *and a craftsman shapes it with his chisel.*
> *They adorn it with silver and gold; . . .*

> *. . . But God made the Earth by his power; he founded the world by his wisdom*
> *and stretched out the heavens by his understanding.*
> *When he thunders, the waters in the heavens roar;*
> *he makes clouds rise from the ends of the Earth.*
> *He sends lightning with the rain and brings out the wind from his storehouses.*

Did those last five lines read somewhat familiarly? Now focused within the space of a single breath are the two aspects of creation we have identified in the Wisdom and Psalm accounts: God's setting out of foundations and boundaries of the world, and its dynamic play caught between harmony (the cycle of clouds rising and rain falling) and chaos (here represented by lightning and wind). Wisdom appears yet again at the foundational stage, and, as in the Psalms, an unstated question seems to hover around the place of creation's more chaotic elements. The other edge of the picture frame Jeremiah constructs for his creation story follows with an immediate return to the theme of idolatry:

> *Everyone is senseless and without knowledge; every goldsmith is shamed by his idols.*
> *His images are a fraud; they have no breath in them.*
> *They are worthless, the objects of mockery; when their judgment comes, they will perish.*
> *He who is the Portion of Jacob is not like these, for he is the Maker of all things,*
> *including Israel, the tribe of his inheritance –*
> *the Lord almighty is his name.*

We notice, by repetition, something surprising about Jeremiah's assault on idolatry: here he attacks directly neither idols nor idolatrous worship, but both before and after his creation story he rails instead at the idol *makers*. Turning pieces of the physical world—wood, stone and metal—into things worshipped is somehow the wrong way of relating to creation. It is a perversion of the Psalmists 'going out to labour until evening' and a betrayal of the wisdom that ought to mark all works of creation, both original ('the Maker of all things') and derivatory ('His images are a fraud'). This is the prophet's development of Genesis' famous *imago Dei* ('His image') notion of humanity. As God creates so, with breath-taking audacity, must the people who reflect His own nature, not to conjure up the false deities of primitivism but to participate in the continuous task of bringing nature itself to fruitfulness.

Prophetic criticism in the Bible is designed not to foretell inevitable disaster, but to portray the consequences of human behaviour in a way that brings about change. Jeremiah recruits creation stories to serve this function, and in a remarkable way that goes unnoticed by modern readers unless we accustom ourselves to the thread of creation tradition we have been following. What should we make of this vision (from chapter 4)?

I looked at the Earth, and it was formless and empty;
and at the heavens, and their light was gone.
I looked at the mountains, and they were quaking;
all the hills were swaying.
I looked, and there were no people; every bird in the sky had flown away.
I looked, and the fruitful land was a desert; all its towns lay in ruins before the Lord
before his fierce anger.

The ingredients of the simple creation sequence are there, but it is as if someone put the film though the projector in reverse. All the familiar elements of the biblical creation story are presented, from the foundations of the Earth to the heavens above, but now in dissolution rather than constitution—this is an *anti-creation* story.[3] Now the boundaries have gone, for there is no longer any form to the Earth. The mountains are no more set in place but in motion, and the fruitfulness of the Earth is set in reverse until all that is left is an empty barrenness. Without having to spell it out, Jeremiah's haunting message is simply that humankind and the physical world are so closely knit that replacing wisdom by foolishness within this relationship can cause creation to run backwards. The cosmos, once so intricately knitted together, unravels before the prophet's eyes.

For a close contemporary parallel to how this prophetic practice works we might think of the film *An Inconvenient Truth*. Aimed as a stark warning of the consequences for the environment and climate of continued thoughtless emission of greenhouse gases by mankind's unbridled industrial lifestyle, it aimed to show not a future that must be but one that may be, depending on our reaction to its message. In a double parallel to the creation themes within the Old Testament prophets, it draws not only on their technique but also on their content of the

[3] There is a strong linguistic clue here in the original text: only here and in Genesis 1:1 does the Hebrew *tohu wa-bohu* occur, translated variously as 'chaos' or 'formless'.

intimate balance between humankind and the physical world. This is a theme already worked out by an earlier tradition of prophecies collected under the name *Isaiah*.

Creation and Care: Later Isaiah

Already woven into the Wisdom and Prophetic texts we have encountered is not only a contemplative relation towards creation but also an active engagement with it. Beginning with the playfulness of the child Wisdom in Proverbs, this idea matures into the fruitfulness of the farmer in Psalm 104. Isaiah is a more political book, its first section (chapters 1–39) immersed in the immediacy of the events leading up to Judah's exile to Babylon, and the following chapters in tracing any strands of hope that might lead to a return. We might have expected that political themes would call attention away from the physical world. After all, we are accustomed to a fragmented media that habitually divorces 'political, social and economic' discussion from scientific. But this is not true of the biblical material, for the responsibility of humans to care for the world in all these realms is rooted in the creation itself. Take, for example, this passage, an expanded version of the psalmist's iconographic farmer, in Isaiah 28, set among criticisms of godless and debauched religious festivals:

Listen and hear my voice; pay attention to what I say.
When a farmer ploughs for planting, does he plough continually?
Does he keep on breaking up and harrowing the soil?
When he has levelled the surface, does he not sow caraway and scatter cumin?
Does he not plant wheat in its place, barley in its plot, and spelt in its field?
His God instructs him and teaches him the right way.
Caraway is not threshed with a sledge, nor is a cartwheel rolled over cumin;
Caraway is beaten out with a rod, and cumin with a stick.
Grain must be ground to make bread; so one does not go on threshing it for ever.
Though he drives the wheels of his threshing-cart over it, his horses do not grind it.
All this comes from the Lord Almighty, wonderful in counsel and magnificent in wisdom.

Much more an invitation to explore technical details of farming than a bucolic reflection of idyllic life on the land, a reader is drawn in to question. It reminds us that, to be fruitful, our working relationship with nature has to be a highly studied one. Bread-making is not an innate skill, but a technology to be learned. Ploughing is an essential preliminary to planting, which itself must recognise soil type and environment fit for each seed. After harvest is the threshing, then the grinding, using appropriate tools for each action. The way to cultivate a fruitful relationship with nature is not obvious or innate. Rather it requires careful study, an activity identified in Proverbs as a part of wisdom. The context of the passage works both ways within Isaiah's narrative: the social stability for which an agricultural people longs can be established only within a nation that itself exercises wisdom rather than foolishness, while a healthy relationship with God endows the nation with the wisdom to work with the natural order.

As the slow crescendo of prophecy builds throughout the Isaiah material from current woes to future hope, so the nature language enlarges its span. The movement culminates in the great extended announcement of restoration and healing for Israel that begins in chapter 40 (so memorably introduced by a piercing tenor voice in Handel's *Messiah*—'Comfort ye, comfort ye my people'), matched by a new natural backdrop. But now it is not the tilled and sown field but the equally measured but infinitely grander cosmos itself (v.12ff.)

> *Who has measured the waters in the hollow of his hand,*
> *or with the breadth of his hand marked off the heavens?*
> *Who has held the dust of the Earth in a basket, or weighed the mountains on the scales*
> *and the hills in a balance?*
> *Who has understood the mind of the Lord,*
> *or instructed him as his counsellor?*

There are two new ideas appearing in this creation text strongly for the first time, which will become increasingly significant for our tracing of structure within ancient narratives of nature. The first is the theme of *number*. A similar attention to detail is there as in the farmer's selection of seeds in chapter 28, but now it takes the form of quantitative measure. Of course the extraordinary idea of God measuring the dimensions of the universe with his hand, and its weight with his equivalent of market scales, serves the immediate purpose of recalibrating the scale of the readers' imaginations. But it also introduces the idea of quantitative measure as an integral part of nature wisdom. It is an aspect of the 'mind of the Lord' to seek out the numerical in relation to the physical world. It is a tiny seed of an idea at this point, but it will grow and multiply.

The other new theme arises more from the context of this creation hymn than its content. For the section of the book that begins at Isaiah 40 is one of the most central in the Old Testament in regard to the 'salvation' (or 'healing'—the Hebrew word *yeshuwah* is employed for both). There has been a tradition of classifying biblical literature into 'salvation theology' and 'creation theology', ascribing different sources to each.[4] Behind this assumption lies a sort of Cartesian 'mind–body' dualism that approaches matters of 'faith' as disjoint from physical substance. Yet here, and in many places to come, we see reference to the structure of the physical cosmos entwined within texts that deal with the longing within Israel to put right what is wrong. Creation theology and salvation theology begin to blur together into a single image. Or, in musical terms, the tenor line in *Messiah* describing in overtly illustrative terms the levelling of mountains and valleys is joined by the chorus announcing at last the revelation of the glory of God, now hidden, before all peoples.

Isaiah returns to a creation theme in the later more explicitly political passages on the international shifts of power (specifically the annexation of Babylon by Persia under Cyrus) that will lead to a return from exile to Jerusalem. From chapter 45:

> *This is what the Lord says – the Holy One of Israel, and its maker:*
> *Concerning things to come, do you question me about my children,*
> *or give orders about the works of my hands?*
> *It is I who made the Earth and created mankind upon it.*
> *My own hands stretched out the heavens;*
> *I marshalled their starry hosts.*
> *I will raise up Cyrus in my righteousness:*
> *I will make his ways straight.*
> *He will rebuild my city and set my exiles free,*
> *But not for a price or reward, says the Lord Almighty.*

Observe how the spotlight on Isaiah's backdrop of creation has panned from picking out the details of fields and harvest on the Earth to the starry hosts of the skies, as the foreground scene changes from public ritual and worship to the playing out of international politics.

[4] The development of creation theological thinking from these passages in the twentieth-century Old Testament scholar Gerhard Von Rad is discussed in: Stefan Paars, *Creation and Judgement: Creation Texts in Some Eighth Century Prophets*. Leiden: Brill, 2003.

The extended creation story of boundaries in Earth and sea, then the emergence of life and the essential presence of conscious humankind within it, is here concertinaed into a single sentence, before our eyes are brought to rest on the stellar backdrop in all its dynamic complexity. There is here even a hint that the motions of the stars follow the orders of their creator.

The forestage character is Cyrus, the Persian ruler who in 539 BCE annexed Babylonia and subsequently allowed a repatriation of some of the exiled Jews to Jerusalem. Again the retelling of the creation story is to practical purpose, but rather than leading towards an engagement in wisdom, or a rejection of idolatry, or even a gentle technology, this time the point is that creation is foundational for the destiny of peoples, and in particular for Old Testament Israel. The first quiet linking of creation and salvation, in its political form, that chapter 40 introduced as a single voice has now become a theme of such a volume that no one could miss it. Much later, the use of astronomical phenomena as the standard metaphor for political events, especially disruptive ones, won permanent adoption in the literary form of *apocalyptic* (the Revelation to St John is probably the best known example, but there are many others outside the Bible from the early centuries BCE and CE).

A Distant Hope and a Different Cosmos: Early Isaiah and Hosea

In the prophetic writings to Israel in exile or earlier, including the first part of Isaiah, and a much shorter book of warnings against idolatry, Hosea, another theme is introduced that plays a sort of counterpoint to those we have heard from already. Equally rooted in the experience of pain and the current experience of both the power and precariousness of creation as in Psalms, Proverbs and Jeremiah, this song responds in a different way to the 'rolling up of the sleeves' of Isaiah 28. Instead of the doer we have the dreamer—a gazing into the far distance of a remote possibility, of another way of ordering the creation. Isaiah 11, a famous 'messianic' passage often read at Christmas services (and, according to Luke's gospel, read self-referentially by Jesus), speaks of the eventual coming of a righteous ruler, possessed of knowledge and wisdom, and through whom justice for the poor and the oppressed will finally arrive.

But the wave of righting wrongs will not be constrained to the boundaries of human society:

The wolf will live with the lamb, the leopard will lie down with the goat,
the calf and the lion and the yearling together; and a little child will lead them.
The cow will feed with the bear, their young will lie down together,
and the lion will eat straw like the ox.
The infant will play near the hole of the cobra, and the young child put his hand into the viper's nest.
They will neither harm nor destroy on all my holy mountain,
for the Earth will be full of the knowledge of the Lord as the waters cover the sea.

This is a created order without prey and predator, and in a very different relationship with humankind. The threatening agents in the non-human world are gone, not we note by annihilating the creatures that currently embody them, but by transforming both them and us. Intriguingly the cause of such a renewed relationship of harmony within creation is identified with a renewed relationship of knowledge of its creator. But is this a vision of knowledge about creation itself (as a participation in what any creator knows and understands of what he or she has made) or a knowledge of the creator? The ambiguity is undecidable.

A similar 'spilling over' of renewal from human to physical creation appears in Hosea chapter 2. Hosea uses a powerful and beautiful metaphor: that of God as the spurned lover longing for the day when his beloved (Israel currently enamoured with other 'deities') returns to him. In a famous subplot, Hosea himself is rejected by his own wife in a brilliant stroke of living metaphor that communicates to the reader the emotional pain of a broken relationship as much as the theological necessity of reconciliation. Through the prophet's voice, God dreams of the day of healing when:

. . . she will sing as in the days of her youth, as in the day she came up out of Egypt.
'In that day', declares the Lord, 'you will call me "my husband";
you will no longer call me "my master" . . .
In that day I will make a covenant for them with the beasts of the field and the birds of the air
and the creatures that move along the ground . . .
I will respond to the skies, and they will respond to the Earth;
and the Earth will respond to the grain, the new wine and oil.

The most powerful notion here is 'covenant'—this is the Old Testament's highest status of relationship. A covenant between rulers was one that openly invited retribution by death were it ever to be broken. It is the eternal promise of commitment of allies, and one that God entered into unilaterally (and uniquely so) with Abraham and his descendants (the foundational account is in Genesis chapter 15). But here the covenant is envisaged with the animal and physical creation itself, the beginnings of a new created order of fruitfulness and an end to strife. It is the same dream as Isaiah 11 but cast in different language, and makes as little sense to those who know that the world is simply not like that, and who believe that there is no conceivable path by which it could become so. But dreams have a habit of recurring, and we will come across this quiet, but extreme, hope given voice to again.

The Formal Tradition: Genesis 1 and 2

One of the many problems facing the reader of the creation stories at the beginning of the book Genesis is that they occupy, by tradition, the first pages of our Bibles. They are, of course, fundamentally important theological and narrative texts, and open up the story of the covenant relation between a creator God and his people. They constitute, however, neither the earliest nor the most fundamental of Old Testament writing on the natural world. As we have seen, they are most certainly not the only 'creation stories'. We must wonder whether much of the deplorable literalism around Genesis 1, still so shrill in many frightened Christian churches today, might have arisen purely as a result of the traditional ordering of biblical books. It takes a while to get to Proverbs and Isaiah if you start reading from page 1. William Brown, in his survey of the seven traditions of creation stories he identifies in the Old Testament,[5] has written extensively on the Genesis stories from the perspective of a scientifically trained theologian. We need not recapitulate his detailed study here, but should distil the outlines of a reading from the perspective of 'wisdom to do with natural

[5] William P. Brown, *The Seven Pillars of Creation: The Bible, Science, and the Ecology of Wonder*. New York: Oxford University Press, 2010. Brown's seven pillars correspond to, and expand on, the different creation traditions covered in the sections of this chapter, developing their theology under a scientifically informed eye.

things', especially one made *after* reading the other major Old Testament creation narratives.

There are two entirely separate creation narratives given sequentially at the outset of Genesis.[6] The first, 'six day', narrative takes all of chapter 1 and the first three verses of chapter 2. Much has been written on the unfortunate and artificial chapter break, which suggests a reading of the first story in which the creation of humankind is the final masterstroke, rather than the actual climax of the narrative which takes the form of the sabbatical, seventh day of rest. At the very least it lends distortion to the meaning of the mandate given to man and woman to 'subdue' the Earth (v.28). The second narrative begins in chapter 2:4, and leads into the events set in the Garden of Eden.

One way to appreciate the distinction between the two stories is as an amplification and decoration of the two powers of 'word' and 'breath' that combined in the creation narrative of Psalm 33. In the psalm the thought moved from one to the other within a short verse—in Genesis the chapter 1 narrative expands entirely on the creative force of Word. 'And God said, "Let there be . . . "' is the formula that announces the coming into being of light, waters, dry ground, vegetation, celestial bodies, animals aquatic and terrestrial, and finally humankind. The chapter 2 account is quite different—here God takes the dust of the ground, watered by streams, then (2:7):

> *The Lord God formed the man from the dust of the ground and breathed into his nostrils the breath of life, and the man became a living being.*

This is much more intimate language, an indwelling breath, the picture of a cradled mass of inanimate matter receiving for the first time the fire of creative, personal love, and assuming life. The narrative form is also more personal, closer and less formal. This second account needs to adopt such a storyline because it will develop—through the scenes in the garden of temptation, the dealing in lies and half-truths and eventual disobedience of Adam and Eve that begins the great biblical story of covenant and reconciliation. We have noticed time and again in the Bible that human relations with nature are couched in terms of pain. In the Genesis 2 narrative this convergence is at its sharpest. The destiny of

[6] Chapter 1 is attributed to the "priestly" source, and chapter 2 to the "Yahwist".

the man formed from the ground is no other than to return to it (2:17ff.):

> *Cursed is the ground because of you; through painful toil you will eat of it all the days of your life.*
> *It will produce thorns and thistles for you, and you will eat the plants of the field.*
> *By the sweat of your brow you will eat your food until you return to the ground,*
> *since from it you were taken; for dust you are and to dust you will return.*

As Brown[7] and others before him observe, only a superficial reading sees in this, and the parallel passages of consequence for the Serpent and for the woman, a simple dealing out of punishment. It is more a grounding of the present predicament of humankind within the history and fabric of the material world—it begins, but does not finish—an *explanation* of the relationship of toil and pain humans have with the material world. By virtue of its gritty truthfulness, it also somehow avoids pessimism. For better as well as for worse it says that our story is bound up with that of nature.

The parallel passage in Genesis 1, to which we have already referred, is initially sunnier in its outlook (1:28ff.):

> *God blessed them and said to them, 'Be fruitful and increase in number, fill the Earth and subdue it. Rule over the fish of the sea and the birds of the air and over every living creature that moves in the ground'.*

It is part of the account of the sixth 'day' of seven, each introduced formally by the invocation 'Then God said . . .' and ending 'and there was evening and there was morning—the nth day'. Although there is progression in the chapter 1 account, it does not lead into any other story. Rather it exists, framed by its formal structure, as do set pieces of liturgy or ceremony today. Brown and independently the Orthodox scholar Margaret Barker both suggest a structural parallelism of the Genesis 1 text with the architecture of the temple, but, whether this suggestion can be sustained or not, what the 'priestly' account does is surely to enshrine the purpose and nature of creation within the repeated acts of worship of the community. The days of the week may well be literal, but if so then they are the repeated days of a cycle of worship and prayer, not a physical account of the beginning of the universe.

[7] William P. Brown, *The Seven Pillars of Creation: The Bible, Science, and the Ecology of Wonder*. New York: Oxford University Press, 2010. Margaret Barker, *Creation*, London: T&T Clarke, 2010.

Seen this way, just as in Genesis 2 the *context* provided an anchor line onto the meaning, so, in Genesis 1, a context of communal remembrance and worship provides the grounding of the text that the lack of a continuous history fails to. In this light, the mandate for fruitfulness is no longer quite so sunny. Taken into the regular reflections of a community subject to droughts and famines, wild animals and epidemics, it reads rather more earthily than as a pure creation fable. Like the Genesis 2 account, it draws on the fundamental creation story material we have seen more simply presented—drawing boundaries between Heaven and Earth, land and sea—but now digs that tradition deep into liturgy, just as Genesis 2 rooted it into history. It also enshrines into the memory of tradition the idea that 'in the beginning' there was, not nothing, but chaos—the *tohu wa-bohu* only appearing here and in Jeremiah's terrifying vision of anti-creation. The closeness of chaos, the sense that the cosmos lives 'on the edge' of disorder, is going to be another significant theme. It characterises the relationship that humankind needs to maintain with our world.

To conclude, we have found in even a very brief survey of Old Testament literature that various versions of the story of creation appear at almost every turn. Far from being confined to an introductory preface to the biblical history of God's developing relationship with people, creation is retold time and again as wisdom, prophecy and history mature. The story is emphatically physical and structural, with an emphasis on ordering, heterogeneity and the establishment of boundaries between Earth, sea and sky. Furthermore it always places creation in a dual relationship with both creator and humankind. It is told to a purpose, developing the relation God's people enjoy with the physical creation in husbandry, worship or politics.

But above Isaiah and Jeremiah, above the poetic forms in Psalms or the formality of Genesis, above even the stunning beauty of playful creation in Proverbs 8, there is one Old Testament book whose towering account of the physical world demands our detailed attention. That book tells the story of a man called Job, pushed to the limits of his endurance by the terrible painfulness of nature apparently out of control. We need to spend considerable time with this achingly beautiful argument over the injustice of inhuman nature. But before we visit the climactic text of the book of Job, we must listen to a few more 'stories of science' from our own times that resonate with its chief subject, so that the ancient and modern narratives of nature are given space to talk to each other. The subject is one we have already noticed as we passed by Psalm 104, Jeremiah and Genesis: the ever-present tension of order and chaos. How does the story we call science come to terms with the chaotic and unpredictable today?

4

Order and Chaos: the Comet, the Storm and the Earthquake

People have always been alert to nature's regularity and rhythm. The cycles of morning and evening, of seasons, of birth and death, are as natural to all cultures today as they were to the Middle Eastern civilisation that enshrined them in the Psalms. Even the much more complex 'hidden' regularity of solar and lunar eclipses was known to the ancient Babylonians and Egyptians, together with sophisticated mathematical apparatus for predicting them. As we saw in the last chapter, these ancient wisdoms together describe a pattern of searching, its goal the comprehension of order in the physical creation.

We should not therefore be surprised to find, in constant tension to this, and running throughout history, signs of affront when our natural predilection for rhythm goes unsatisfied. In particular we find that human cultures are especially sensitive to those recalcitrant systems of disorder and inherent unpredictability that take their place in both heavenly and earthly creation. If the road to understanding order and regularity in nature is a long one, how much more tortuous is the search for an intellectual grasp of disorder and chaos. This chapter sets us down at a few viewpoints on that journey, in which we are discovering that science is, perhaps unexpectedly, as open to the unpredictable as it is to the predictable. We will find that the essential act of scientific wisdom is to ask the right question. It will act as a preliminary 'reader' in how we currently think about chaotic systems in nature—we need to experience a taste of this challenging and beautiful realm of current science before we explore in depth the classic Old Testament text—the book of Job—that confronts uncontrolled chaos in nature with the greatest felt indignation.

This chapter also attempts to explore what 'understanding' might mean in the context of chaotic objects such as clouds, storms and earthquakes. We will use metaphor and diagram rather than explicit mathematics (doubtless to readers' relief) but, as in some of the stories of science in Chapter 2, will attempt to engage in that uniquely human story that takes us beneath the surface of nature into the structures and causes that lie beneath. We need to participate in this mental journey of 'wisdom about natural things' before we can decide whether it meets our cry for understanding over ignorance.

The Comet

Eclipses provide us with an interesting example of how understanding can overcome fear. In cultures where their inherent pattern was unknown, and to whom they appeared to be a random disordering of the heavens, there is evidence that they could cause profound terror. But this could not of course be the case for people whose astronomers were able to predict them. No ancient civilisation, as far as we know, however, came close to reconciling an ordered and regular model of the heavens with the awkward and sporadic appearances of the silent and silvery visitors we call comets. I remember the beautiful sight of comet Hale–Bopp, visible in the northern winter sky of 1995, with its bright nucleus and splendid sweeping tail. But I also recall my surprise at finding in my own response an inkling of the *fear* that comets were able to elicit: there it was bright among the familiar northern constellations of the Great Bear and the Dragon, but somehow unbidden and out of place. The sky was just not supposed to look like that! Even the heavens—the home of the ancient Greeks' perfect spheres and circular motions—is also apparently the habitation of irregular and chaotic visitations. So unwelcome was such a conclusion, that Aristotle and his Islamic and Christian commentators in the Middle Ages concluded that comets could not be celestial phenomena at all, but in actuality vapours confined to the ('sub-lunar') space below the spheres of perfection. Comets thread their way through the history of science as travellers on the boundaries of the opposing regimes of order and chaos, but in so doing serve as important pointers, not to disastrous future events but to the social role that science has played in demystifying the heavens.

Might the apparent irregularity of comets, like eclipses, be revealed as a hidden kind of order? There is recorded an amusing but extremely profound moment in the history of science in the London of the late seventeenth century. Edmund Halley, Astronomer Royal, sought out Isaac Newton in 1684 with an urgent question that he realised was beyond his own powers to answer. Halley had made a detailed study of the historical observations of comets, and had detected regular periodicities in some sightings. Could it be that not only planets followed elliptical orbits around the Sun, as Kepler had deduced a century before, but comets as well? If so then some of the regular appearances of comets would be explained simply as repeated visits of the *same* comet. Realising that, if true, there must be an underlying law or common cause behind such 'universal', or shared, behaviour, he had attempted to deduce what its mathematical form must be, but without success. If anyone living were capable of such a challenge, that person was Isaac Newton. Halley was right: 'I have calculated it—the notes are here somewhere' was the gist of Newton's reply, accompanied by a riffling through stacks of papers from which would later emerge his magisterial and timeless *Principia Mathematica*. But on that occasion to no avail—all Newton could recover was the memory of a calculation, but for him the penalty was just the time it took to make a recalculation before Halley had his answer: if the gravitational force pulling a planet or comet towards its central star fell away with the *square* of the distance from the star, then the resulting orbit would take the form of an ellipse. Halley later funded the publication of Newton's *Principia*, containing this and other calculations in the mechanics of gravitation.[1]

It is one of the most magnificent expressions of the mathematical development in physics that had begun four centuries before with Robert Grosseteste and Roger Bacon. Because mathematics is not simply descriptive but *functional*, because mathematical equations take on a life of their own and capture a sort of 'working model' of the external systems they describe, they can be used to answer questions in the same way that building a scale model of an architectural design can predict the stability and sightlines of the real building, albeit in miniature. What Newton was able to do for the first time was to set such a mathematical model running, feeding into it a specific assumption about the structure of gravitational pull from the Sun onto a planet or comet. That very special assumption can be described intuitively: Newton supposed that the weighty tug of gravity was directed always towards the Sun, becoming stronger as the planet approached, and feebler as it distanced itself from the Sun. Furthermore, the variation in force followed the

[1] See Richard S. Westfall, *Never at Rest: A Biography of Isaac Newton*. Cambridge: Cambridge University Press, 1980.

same quantitative proportion as the intensity of the Sun's light increases with proximity and reduces with distance.[2] His mathematics (specifically the then new techniques of the 'calculus') enabled that law, operating at each point of an orbit, to be extended to calculate its entire shape. More than that, the little mathematical machine would also predict the future pace at which the orbit is followed.

The result is surprising in elegance: the curved path turned out not to be some strange new shape, but the special form of an elongated closed curve called an 'ellipse', well known to the mathematicians of ancient Greece. Imagine a conical hat resting on a table, then chopping off its point with a clean slice through the cone, but taken at an angle to the horizontal—the lenticular curve of the cut will have the form of an ellipse. The more closely horizontal the cut, the more the shape approaches the circle; the closer the cut comes to the angle of the cone's sides, the more elongated the ellipse. All these shapes were possible orbits in our (or any) solar system, the actual shape depending on the history of formation of the orbits. Planets tended to adopt near circles, comets (at least those which regularly returned to the environs of the inner planets) more extreme elongations (Figure 4.1). Furthermore, the execution of the orbits would be regular. After one entire revolution, the planet would be back to its starting point, with precisely the same velocity as it had on its previous visitation. Those conditions together are enough to ensure that each orbit follows a history identical to the last, and deliver a perfectly regular periodicity and predictability. Ancient civilisations from Egyptian to Mayan had noticed and exploited astronomical regularity in the prediction of eclipses, but now their inner workings had been exposed.

The mechanical view of the solar system became embodied in the literal working models of 'orreries'. Hidden gears within the orrery table transfer the rotational motion of a turned handle in perfect proportion to each small brass sphere representing a planet. Some even contained smaller and finer gearing so that tiny moons orbited the planets as they themselves waltzed around the central solar orb. These beautiful working models of the solar system in brass clockwork, increasingly popular in the eighteenth century, reinforced in concrete terms the claims of the famous French mathematician Pierre Simon de

[2] This is the 'inverse-square' law: a planet twice as distant from the Sun as another is subjected to only a quarter ($1/2^2$) of the force.

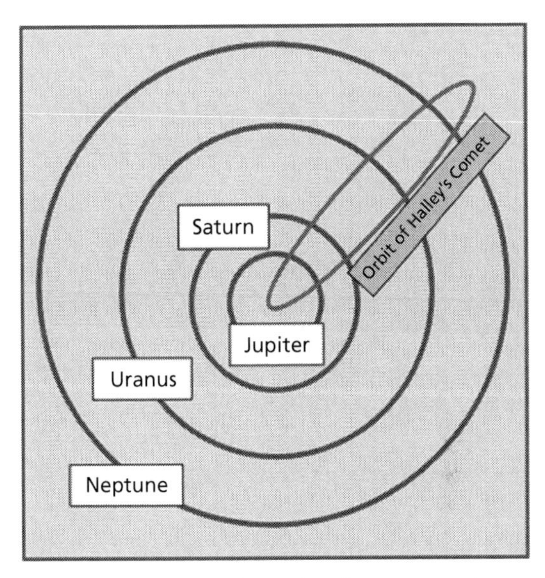

Figure 4.1 The highly elliptical orbit of Halley's comet shown against the orbits of the planets.

Laplace. He dreamed that a being with perfect knowledge of the original positions and velocities of all particles making up the universe would in principle be in perfect possession of all future behaviour.[3] Simply turn the handle on the giant orrery of all matter and you find out what it does for all times into the future.

An overwhelming impression of the massive reality of the planets and moons swinging inexorably in their courses came to me at the moment of a total solar eclipse visible from the south-west of England in the summer of 1999. I say 'visible', but unsurprisingly from where our young family was standing on a cliff top overlooking the sea to the west, our country's habitual cloud cover hid all view of the Moon actually passing over and covering the solar disc. But it gave us something else: the cloud was relatively thin and high, so our experience was instead one of the *shadow of the Moon*, appearing as a dark scar on the distant westerly horizon then rushing towards us at thousands of miles per hour before cloaking the landscape around in 4 minutes of darkness until

[3] Pierre Simon de Laplace, *A Philosophical Essay on Probabilities*. Transl. F.W. Truscott and F.L. Emory. New York: Dover Publications, 1951.

brightness followed once more. The experience seemed almost *audible*: an impression of the sound of cosmic clockwork, a sort of deep rumbling, as an equally imagined geometrical arm radiating from the Sun to the Moon then onwards to the Earth swung over us. We were inside the real orrery of the solar system.

The regularity and predictability of clocks, natural or man-made, is impressive, but is it typical? Is Laplace's vision really the right way of looking at the universe? Are Newton's mathematical laws the last, or the first, word in our rewriting of the book of the world? Even keeping our gaze skywards tells us rapidly that all is not like this. Follow the shape of a passing cloud: great plumes of white opacity bubble up from its heart, creating those fabled imaginary shapes of childhood. Witness the more violent display of a thunderstorm and the tense uncertainty of the next lightning strike, then its jagged undirected path. These events bear no resemblance to the regular pass of the planets, impressive though that is. Under wider examination, the world seems balanced between phenomena of order and disorder, of regularity and bewildering chaos. We can even identify examples of phenomena sitting on the knife-edge point of balance between the two: after blowing out a candle watch for a moment the thin wisp of smoke as it rises from the smouldering wick. At first its rising path is smooth and its shape a gently necked cylinder, directed vertically upwards and thinning as it accelerates under the buoyant force of the cooler air around. Then suddenly it breaks up into whorls and knots, twisting and breaking in a turbulent, chaotic writhing that defies prediction. Or sit, as did a fascinated Leonardo de Vinci, by the side of a stream and watch as the smooth flow of water approaching a rock breaks into frothy whorls of white-water as it passes (a beautiful example is reproduced in Figure 4.2).

Even the most powerful computer codes working with the laws of fluid motion cannot anticipate the exact form of this unstable 'turbulent' sort of fluid flow, the predictions beginning to diverge from reality very soon after the smooth form finishes. Why is it that some parts of nature behave regularly and others apparently randomly? And is the randomness real or illusory? Could we in principle just build a bigger computer and compute turbulent flow after all? How much of the dynamic world falls into each class? These are important practical questions—just as the lightning strikes the Earth seemingly at random from above, so earthquakes from below remain infamously out of reach of prediction—and this in spite of monitoring equipment, and

Figure 4.2 A sketch from Leonardo da Vinci's 'Study of water falling into still water', c.1508–9. Royal Collection Trust/© Her Majesty Queen Elizabeth II 2013.

mathematical models of the Earth's crust based on detailed soundings near fault zones. Many lives would be saved if we understood and could predict the subtle dynamics of stress and slippage of rock deep within the Earth's crust. We do better with the prediction of hurricanes and storms from above, but then only for a few days ahead. These ubiquitous phenomena also tug at our deep desire to understand the world. If we cannot capture a form of comprehension by a predictive science, then by what other route might we create understanding?

The answer to the question about the prevalence of order over chaos turns out to be bad news if your predilection is for universal tidiness: smooth predictability is the preserve of special and small havens of calm in a giant ocean of storms. We are therefore not going to set our sights at an understanding of every swirl and eddy of chaotic motion. Instead we will begin with the lesser goal of an insight into *why* chaotic motion occurs, and *what* aspects of it we might hope to comprehend. Even this is by no means a trivial task, and begs questions along the way of what we might mean by 'understanding'. Our tools will be intuitive rather than mathematical: we will explore a simple but unfamiliar chaotic system by making correspondence with another situation more easily grasped by experience and insight. A tried and tested way of explanation is to map the unfamiliar to the familiar—in this way the islands of knowledge slowly grow into the surrounding seas of ignorance.

We start not with a chaotic system, but a regular one which still contains a surprise. It will also provide us with an invaluable tool—we will be able to understand the system better by mapping it to an equivalent one that we think of in a different way. It is the simplest dynamical object I can think of, whose regularity set the standard for timekeeping for centuries: the pendulum. As a thought experiment, imagine hauling the bob over to one side of its swing, waiting until it becomes completely still, then letting go. We know what happens—picking up speed as it descends towards the midpoint of its arc, it then climbs to a near-equal height on the other side, slowing as it goes then coming briefly to a stop, before turning and repeating a mirror image of its first swing. It returns to its starting point (almost), thence to begin again, and again, ... Surprisingly, when the angle of swing is small the time to execute a complete cycle of the motion is independent of the size of the arc—but a moment's thought perceives the reason. The larger the swing, the larger the speed it builds up towards its lowest point. Providing the two are proportional, the time taken for any swing will be the same. If at each corresponding point of a swing of twice the span as another, the speed is also twice as great, then the two trajectories will take precisely the same time to sweep out. Of course it is just this stability of period with amplitude that makes the pendulum an ideal timekeeper in a clock (if you have room for it).

The gentle rocking to and fro is the most common example of the physicists' 'simple harmonic motion'—the same pattern, but executed much faster, lies at the heart of music in the vibration of simple tones; much slower it is exemplified by the motion of Jupiter's satellites seen in projection tracing and retracing their paths across the planet. The mathematics is now available to eight decimal places on every high school science student's calculator.

But more still is true—now take the pendulum again, but this time rather than simply releasing it give it a mighty push. Now rather than oscillating to and fro the pendulum rotates wildly around its pivot—a little slower near the top of its arc than the bottom, but a motion of quite different form that never reverses. Ringers of English church bells are familiar with the transition between these two forms of pendular motion: too hard a heave on a bell rope will cause a heavy bell to swing right over its vertical position, sometimes with enough energy to break the 'stay'—a thick wooden arm meant to prevent just such continuous revolution—both winding up rope and lifting the hapless bell ringer towards the ceiling. How can we understand this radical divergence of two types of behaviour in the same system?

This is where a mapping to a different picture helps us win some insight. The key idea in the mapping is *energy*—the potential of the swinging bell to break the stay or to accelerate downwards to its greatest speed of rotation. In the presence of gravity, a higher object possesses more energy by virtue of its position than a lower object (we call this *potential energy*). As the object descends, the energy appears in the alternative form of its motion (this is *kinetic energy*). The wonderfully useful property of the energy of a system effectively isolated from its surroundings is that the sum total of all its forms of energy remains constant. As a pendulum descends, it loses potential energy but speeds up, gaining kinetic energy at an equal rate. Providing that frictional losses are small, the pendulum's motion corresponds to a regular transfer and return of its unchanging total energy between these two forms. We can picture what is going on in terms of a 'landscape' of the potential energy. Thinking of the angle traced out by the pendulum from its initial position as a type of distance, we can then draw a map of its potential energy as this distance is traversed. The map looks like a perfectly repeating undulation of hills (Figure 4.3).

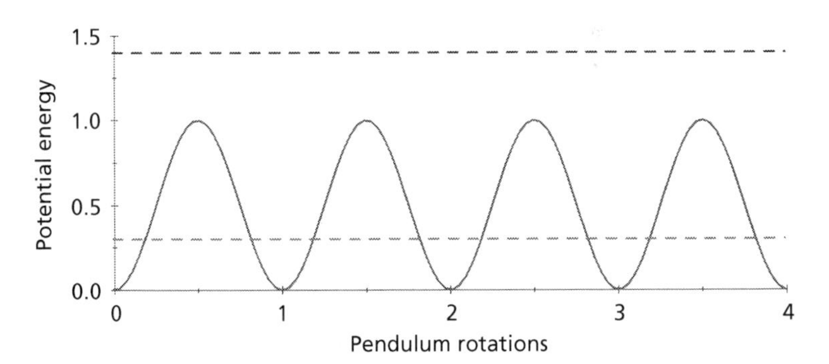

Figure 4.3 The landscape of potential energy of a pendulum swinging out at a continuous angle from left to right. Each repetition of the hilly shape corresponds to a whole rotation of the pendulum. If its total energy is small (level of the lower dashed line) then it is trapped into oscillations within a single valley; if large (level of upper dashed line) then it is free to roller-coast from one valley to the next.

Viewed in this way, the motion of the pendulum is equivalent to that of a roller-coaster car on a track with the shape of the grey curve. Both the roller coaster and the pendulum possess potential energy proportional only to their height from the ground, gaining speed as they descend and losing it as they climb. Suppose the total energy of a particular swing is represented by the lower horizontal dashed line in Figure 4.3 (this could be achieved by letting the pendulum go from any point on the curving landscape at the height of the dashed line). The roller coaster on the track (or the pendulum moving through the equivalent angles) begins to descend downwards to the nearest valley bottom, and to climb up the other side. But it can climb no further than the height at which it began (because of the conservation of energy). Then the pendulum is trapped inside the valley it started in: there is no escape from an oscillation back and forth between the two positions of maximum height, where it meets its ceiling of maximum potential energy. For at this point the kinetic part of its energy is zero. All energy is delivered to the potential form—it can therefore move no further.

Now imagine a much greater total energy, say at a value represented by the higher dashed line in the figure. With this total energy there is plenty to spare (in the form of motion) when the pendulum reaches the peaks of the potential energy landscape. Like a roller coaster beginning from a starting point higher than any subsequent rise, it escapes all of the valleys it meets and traverses the whole landscape. The equivalent pendulum rotates round and round its pivot. We can even see from the roller-coaster landscape picture that there exists a critical initial energy—when the total energy is just enough to make it to the peaks in the landscape—which divides the regimes of motion: oscillating in a single valley for smaller energies and continuous motion in one direction for larger.

More than just general ideas, such as the laws of motion and the shifting of energy between forms as a result of motion and position, connect the pendulum and the comet. Both planets and the periodic comets (like Halley's) pursue closed orbits that recur with predictable periods. Their total energy is insufficient to escape completely from the Sun's gravity, and, like the 'low-energy' pendulum trapped into an oscillating pattern, they remain bounded within the solar system. But there are other comets that visit the neighbourhood of the Sun and Earth only once. Their orbits do not close (following not ellipses, but related open curves called hyperbolae). Just like the freely rotating pendulum of high energy, these comets possess an initial energy of motion sufficient to escape solar gravity completely, swinging by the Sun never to return. With the idea of an 'energy landscape' in our mental toolkit we can explore less well-behaved systems.

The Storm

They begin as benign stirrings of warm air over the South Atlantic. Fed by in-falling winds from surrounding regions of higher air pressure, the rotation of the Earth itself imparts a slow circulation to the growing roundabout of wind. There are no advanced signs of the conception of an infant hurricane, but its growth to maturity is all too obvious, and its relentless path all too menacing. With days' warning at best, winds of terrifying force batter coastal communities, razing communities and severing supply lines. It would be hard to imagine anything in nature more removed from the regular and predictable whirling of the planets. But understanding the storm is the first step towards living with it. Where should we begin?

Although a simple idea, we have made a deep and powerful leap of thought by connecting dynamics to geometry in the roller coaster analogy for the pendulum. It is an idea that dates back at least to the fourteenth-century natural philosophers of Paris and Oxford. The return is significant in terms of insight and prediction. We can 'see' straight away how the system will behave, and with a further connection to algebra can turn this insight into predictive numbers. With imaginative and mathematical tools like these, we begin to understand how the extrapolation of this sort of success towards the capturing of much more complex systems, even the universe as a whole, appeared as a possibility to eighteenth- and nineteenth-century mathematicians (although I do wonder if Laplace had ever made himself think that thought through while standing in the face of a raging winter storm). But as we anticipated from our observation of complex dynamics in the world around us, let alone in the teeth of a hurricane, a surprise awaits, and much sooner and simpler than expected. Suppose we design a more complicated 'compound' pendulum in which the pendulum arm is connected to the bob via a second arm pivoting on the end of the first. This is the smallest additional structure one could imagine adding to the simple pendulum—essentially one pendulum joined onto another. If limited to small swings, it behaves rather as one might expect, with each arm oscillating gently back and forth with its own characteristic period. But if we release the bob from an elevated position, an extraordinary

Figure 4.4 The path of a double pendulum visualised in a long-exposure photograph of a light source attached to the end of the pendulum. Image credit: George Ioannidis/Wikimedia Commons.

pattern of behaviour unfolds. The bob swings, rotates, reverses in a wild and unexpected fashion. Figure 4.4 illustrates, by long exposure of a light at the end of such a 'double pendulum', an example of the, truly chaotic, motion.

The regularity of even the 'high-energy' motion of the simple pendulum has completely disappeared, along with any known mathematical form that might describe it. Playing with this beautiful toy, alongside careful observation, reveals another strange aspect to its behaviour. Suppose one tries to repeat an experiment by releasing the bob for the second run as close as possible to the starting point of the first. We expect the same starting conditions to be followed by the same motion, but time and again, although the second trial starts off performing the same spins and loops as the first, after a short time the pattern starts to deviate until it becomes unrecognisable. The bob executes the same sort of motions in a statistical sense (from the picture we can see that it spends more time making small loops at the bottom of its swing than large excursions to the side, for example), but all exact correlation from one experiment to the next disappears. Can we understand why?

The power of the analogy we have set up between the pendulum and 'roller-coaster landscapes' now comes into its own. As a first step, we might think about what an 'energy landscape' for the double pendulum might look like. We now have two angles, rather than one, that together specify the state of the object, and so the energy (recall this is just the height) of the bob. These may be thought of as just the angles of both connected pendula to the vertical. Increasing either angle on its own will result in the undulating curve of the simple pendulum, so the best way of representing the result of letting both vary is to create a *two-dimensional* landscape that contains both contributions. Suppose we 'unroll' the angle of the first arm of the double-pendulum (as we did to construct the roller-coaster landscape of Figure 4.3) in the east–west direction, and the second north–south. Now every point on our landscape (which begins to look a lot more suitable a term) corresponds to a particular position of both arms. The height of the landscape corresponds to the potential energy—varying from a maximum when both pendula point straight up to a minimum when both are down. Viewing this energy landscape from an oblique angle gives a view such as that shown in Figure 4.5. It is a regular pattern of rounded hills, valleys and saddles.

The motion of the double pendulum corresponds to the motion of a ball rolling about on this landscape in just the same way that the single pendulum corresponded to the simple one-dimensional roller-coaster landscape we discussed before. But now, rather than the roller coaster, kept to its

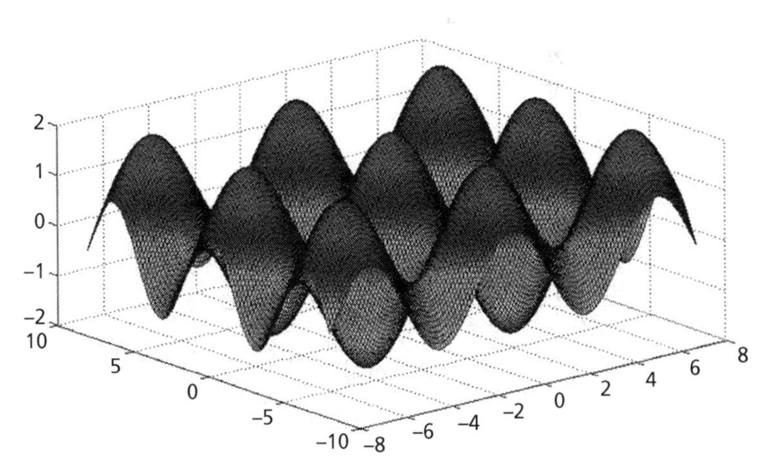

Figure 4.5 The 'roller-coaster' energy landscape of the double pendulum, equivalent to a periodic pattern of hills and valley covering a surface, rather than a line.

tracks and predictable in its path, the addition of the new direction has given us a real golf-player's nightmare of a landscape. Imagine trying a long putt across a green like this! This impossible task is exactly equivalent to the prediction of a double pendulum's motion. The geometrical picture is still powerful, but its power lies not so much in showing the way to a solution but in seeing *why* the solutions are so hard to calculate. Suppose a trajectory on the landscape passes close to one of the regular 'hilltops'. Two slightly different paths, one passing just to the right and the other just to the left of the maximum, will execute very different subsequent routes. One is sent off diverging to the right, the other to the left. Very soon they will be passing through different valleys entirely, although they began as travelling companions. In fact this rapid divergence of initially close paths happens for all initial directions.

We also know from watching tense final shots on real putting greens that the final direction of travel can be very sensitively dependent on the speed of motion as well—as a hill is approached obliquely, slower trajectories are deflected much more strongly that faster ones. With each successive navigation of a rise and fall across the surface, the uncertainties in the future path grow exponentially.

We can also see that this type of trajectory, highly sensitive to its initial position and velocity, only appears at high energy, when the system is able to escape from its initial low-energy valley. The high-energy behaviour in just one dimension (the simple pendulum) was interesting but not problematic from the point of view of computability of the future behaviour. But the twin effects of high energy[4] *and* higher numbers of variables (or equivalently higher 'dimension' of the landscape—even two variables are enough) creates rather naturally the complex behaviour which will diverge from any practical prediction before long. This is one of the features of the phenomenon known both colloquially and technically as 'chaos'. Accessible at many levels—mathematical, physical, recreational and artistic—the field of chaos has, since the early 1980s, spawned a revolution in physics, and a series of books, articles and striking graphic images in the public domain.[5]

First glimpses of chaotic dynamic behaviour appeared in 1961, in a simplified computational model for weather prediction by Edward

[4] The important criterion is technically that the dependence of the forces on the position of the pendulum should be 'non–linear'.

[5] See, for example, James Gleick, *Chaos: Making a New Science*. London: Viking Penguin, 1987; <http://chaos.aip.org/gallery_of_nonlinear_images>.

Lorenz of the Massachusetts Institute of Technology in Boston, MA. Just think a while of how many variables might be linked together in a calculation of the complex systems of wind, temperature, humidity, water and air that we call 'weather'. Even if temperature and wind velocity are sampled only on a large 'grid' of 1-kilometre cubes, a slab of atmosphere 10 kilometres high the size of the UK generates over 20 million coupled variables—a far cry from the already chaotic double pendulum. Lorenz managed to extract a system of just three variables by making huge approximations to a simplified model of one aspect of a weather system: the large circulating vortex tubes of air set off by rising warm air. He noticed immediately that small rounding differences in the numbers fed in as initial conditions to the calculations soon produced widely different solutions. This is the effect we identified in our 'golfer's nightmare' landscape for the pendulum. He also observed other features of chaotic dynamics that proved to be very significant—the 'mixing' property of chaotic flows, for example, describes motion which eventually spreads any initial patch of original values everywhere in the system.

Today's weather prediction recruits the largest available computing power, and a richness of real-time measurement of atmospheric conditions undreamt of by Lorenz. Yet we are all well aware that, for more than 5 days ahead, no one knows what is going to appear in the sky (at least this is true of places with 'real weather', such as the country I live and work in). The storm, emblematic of both challenge and threat from nature, is an example of an intrinsically unpredictable world of rich chaotic phenomena. Paradoxically, we now have some insight into *why* it is unpredictable as well as, and thankfully, some ability to warn of its approach for a short time in advance. But we also understand that our knowledge of this and other chaotic systems will always be partial.

The Earthquake

No one who has experienced a serious Earth tremor is ever again likely to use the expression 'terra firma' as a metaphor for lasting solidity. If the storm symbolises unpredictable terrors from above, then the earthquake remains emblematic of threats to life and livelihood from below. Small wonder that the biblical material refers to the tremblings of the Earth as much as to the storms of the sky to depict the 'inhuman otherness' of nature in all its power. The terrible Lisbon quake of 1755 shook out what optimism had gathered around the scientific progress of Newtonian mechanics, famously in the words of Voltaire,[6] that nature 'might thereby be rendered benign'. It is one of the great frustrations of contemporary Earth science that a theory of earthquakes, let alone a predictive theory, has made so little progress. The current scientific understanding of the processes and patterns has been criticised in recent reviews of the field as handicapped by deep contradictions between theories, and the hopeless task of predicting essentially random events.[7] Partly as a consequence of this, and some spectacularly wrong predictions in the 1970s and 1980s of quiet periods along fault zones, which nevertheless produced quakes causing considerable disruption and damage, the applied field of earthquake prediction has entered an extended period of self-doubt rather than progress. Most shocking of all was the prosecution of six Italian seismologists who failed to predict the severe L'Aquila earthquake of 2009. The civil engineering of buildings resilient to tremors has proceeded with greater success, so that quakes that would in previous generations have caused major collapses now need not do so (all the greater is therefore the outrage over loss of life when this could, by better building, have been avoided). Even so, the few days of storm warning available to us remains out of our grasp in the equally chaotic movements of the Earth.

Our current position is not for want of research, or a long history of it. As we saw in Chapter 2, the desire for understanding and explanation appears and develops throughout the scientific works of the first millennium. Having already met the great seventh-century English scholar Bede, whose *De Natura Rerum* dates from around AD 700, we might consult what he has to say about earthquakes:

> *They say that an earthquake is caused by the wind, which having been shut up in the Earth's cavernous sponge-like innards, rushes through them with a terrifying roar, and labouring to escape, shudders violently with various rumblings, and endeavours to discharge itself by shaking open a gap. Hence hollows in the Earth are associated with these quakes, seeing that they have the capacity for wind, but sandy and solid places lack it.*

[6] Voltaire, *Candide*. London: Penguin, 1997 (first published 1759).

[7] See, for example, Ian Main, Debate: Is the reliable prediction of individual earthquakes a realistic scientific goal?, *Nature* 1999, 25 February. <http://www.nature.com/nature/debates/earthquake/equake_frameset.html>.

The point here is emphatically not to crow over Bede's mistake. He was indeed completely wrong about the cause of earthquakes, which are nothing to do with trapped air. However, he believes that it is worth his time to seek out a physical and mechanistic explanation for the phenomenon. Furthermore, hidden in his account is a logic that points to a deliberate scientific process he is using. Bede employs the (Aristotelian) physics of his time to motivate and guide his search for a cause. For Aristotle, the sub-lunary cosmos was structured through the four elements of earth, water, air and fire, each finding their own natural level. These natural loci followed their densities—hence water, for example, typically lay below air and above earth. Whenever it does not, motion ensues that restores the natural state. Rain would be an example—it arises from water situated *above* air, so falls. Likewise, should air become trapped below earth, this world-view would expect similarly ensuing dynamics—where the structure of the world permits it, the trapped air should seek to escape upwards to restore its natural equilibrium. This is Bede's idea, which he adopts but extends from previous comments of Pliny (in the first century CE) and Isidore of Seville (in the seventh century). Kant revisited the idea, independently as far as we know, following the Lisbon earthquake.[8]

Interestingly, wind theories of earthquakes arose elsewhere in the ancient world, notably in Han Dynasty China. This idea seems to have occurred, again quite independently, within different civilisations in the ancient world that nonetheless shared a tradition of questioning curiosity. Bede observes that it also carries an explanation of why some regions are prone to earthquakes while others are not, speculating on the variable porosity of rocks deep under the Earth's surface. As we saw in Chapter 2, Bede does achieve an example of his stated aim in compiling a *De Natura Rerum*: a theologically motivated set of physical explanations of natural phenomena so that his readers would not impute supernatural causes to them or derive unwarranted fears.

There is a conceptual thread, however, in Bede's thinking that does find a similarity within the rich mix of today's scientific grappling with the complexity of the Earth's crust. He identifies the spontaneous motion of earthquakes as arising from a displacement of matter away from its equilibrium. This idea has a long and fruitful history, developing

[8] O. Reinhardt and D.R. Oldroyd, Kant's theory of earthquakes and volcanic action, *Annals of Science* 1983, **40**, 247–72.

throughout the Middle Ages and Enlightenment in the context of motion and force. As we glimpsed when we visited the statistical mechanics of peptide assembly in Chapter 2, the notion of equilibrium also underpins the science of thermodynamics. In that case, it was the driving force towards equilibrium that generated the change from disperse peptide molecules to the state of self-assembled fibrils.

A recent surge of interest and research takes this idea of balance and equilibrium into the heart of a science much closer to that of earthquakes—the deceptively familiar world of *granular materials*. Think of sandcastles on the beach, weighing out rice on kitchen scales, teasing ground coffee from a packet. William Blake in his poem *Auguries of Innocence* invited his readers 'to see a world in a grain of sand'. For us, however, it is worth spending some time looking at the world inside a whole *pile* of sand, and at some of the beautiful insights that have emerged from this work. Although it does not solve the problem of earthquakes, it opens our eyes to some of the complexities that must lie there, and points to new ways of thinking about them. Just as with the chaotic atmosphere of storms, simplified many million-fold in the gyrations of the double pendulum, simplified systems of granular materials can teach us the wisdom of which questions we should ask, and which we should not.

At first thought, nothing could appear simpler than a stationary pile of sand grains, but look again at the pile on the beach, composed perhaps by lazily dribbling grains onto its centre from your fist. If the pile is relatively flat, the grains just settle where they fall, building the pile slowly upwards. But if the side of the pile is already sloping steeply, the grains will instead slide down it from their landing site. Indeed, close to the angle at which this begins to happen, little avalanches of sand are always occurring. Sometimes these landslides can involve large fractions of the entire slope, especially if there has been a sufficient build-up of the pile without any slippage for a while.

Another thought experiment: consider the hourglass (miniature versions are often used as egg timers). The flow of sand from upper to lower chambers looks superficially like that of any fluid that might do the same—water or oil, for example. But careful inspection discovers an intriguing difference. Water flowing from a container through an exit point towards the bottom will emerge at a variable rate: the greater the 'head' of water, the greater the pressure at the bottom, and the faster the fluid emerges. Sinks and baths empty more quickly at first than when they are nearly drained. But this is not true of sand—the rate at which grains pass into the lower glass chamber (making a tiny version of the sand pile we looked at before) is independent of the amount of sand resting above it. An increased weight of sand in the upper chamber does not result in an increased rate of flow into the lower. This is one reason that sand timers make better clocks than water timers. But the subtle behaviour is not intuitive, and indicates that a pile of loose grains contains hidden structures within it that are different from both normal fluids and elastic solids, even though they are able both to flow like the first, and rest still holding their shape like the second.

Recent research has uncovered something of the hidden world inside a pile of sand, thanks to an international 'conversation' of experimental and theoretical ideas and techniques. Suspicions that granular matter at rest was hiding something interesting arose from the very first attempts to solve the problem of forces between the particles. A typical grain might be in contact with four or five other grains above and below it. With each one it will push with a force of contact (and by Newton's laws be pushed back with the same force). Students of physics are familiar with this sort of problem. We are supposed to be able to work out what these forces are by applying the two simple criteria of mechanical equilibrium: (1) all the forces on each grain from its neighbours must add up perfectly to the single force required to support its weight and (2) there must be no net tendency to rotate or twist the grain. In regular and simple structures the mathematical equations generated by these conditions are sufficient to specify all the forces, but in our sand piles we simply run out of equations before we have found all the forces. All sorts of possible solutions arise for exactly the same pattern of grains. What could such a surprising impasse mean? One consequence is that it ought to be possible to create sand piles of identical shape, but with different patterns of forces within them.

A beautifully simple experiment by experimental physicists in Eric Clément's group at the Université Pierre et Marie Curie in Paris showed starkly just how different these patterns could be.[9] The team created two identically shaped conical sand piles, but by different *histories*. They grew the first from a continuous stream of sand cascading

[9] L. Vanel et al., Memories in sand: experimental tests of construction history on stress distributions under sandpiles, *Physical Review E* 1999, **60**, R5040–3.

onto its apex, like the idle game on the beach. A second construction created an identical shape by depositing carefully each layer of sand grains at a time, starting from the bottom and finishing with the point at the top. Sand grains fell as a uniform 'rain' across the whole of the current pile area, rather than just at the centre point, but this time the experimenters did not allow the sand to cascade down the sides.

Underneath each pile was a specially instrumented base that was able to measure the downward force of weight at each point under the sand pile. Although the outward appearance of the two piles was identical, the distribution of weight underneath them was decidedly not: the pile constructed layer by layer by the uniform 'rain' of particles across the pile recorded, as we might expect, a greater weight under the central peak, falling off towards the outside of the cone. However, the pile created from a single stream pouring down on the centre of the conical pile behaved very differently (Figure 4.6). In this case the weight actually *dips* at the centre—even though the mass of sand above that point is greater than anywhere else. The weight of the central cone is somehow transferred sideways—there is a maximum in weight on a ring some way towards the outside of the circular base. In Figure 4.6 it is clear that the outward appearances of the piles are indistinguishable. Importantly, the angles of slope of the outside of the piles are identical. Clearly something very surprising, and very differently structured, must be going on inside the two piles. And whatever that something is, it must be very far from the structures inside an ordinary 'elastic' solid such as wood, rubber or brick.

Other beautiful experiments in several laboratories have given us the characteristic and essential view 'beneath the surface' of phenomena that opens up the scientific imagination. The technique here is to look first not at three-dimensional piles (this is very difficult to do) but at two-dimensional versions composed of disc-like transparent particles in thin slice-shaped piles trapped between transparent walls. The small discs are made of a material that modifies the light passing through them, depending on the force they are carrying. Forming an image with the light then allows the experimenter to see where the internal stresses within the pile are high and low. A simple elastic network of such particles would show the force gradually building up from the top to the bottom of the pile as the weight bearing down from above increases. Something very different is actually observed. In Figure 4.7 we show results from the Manchester

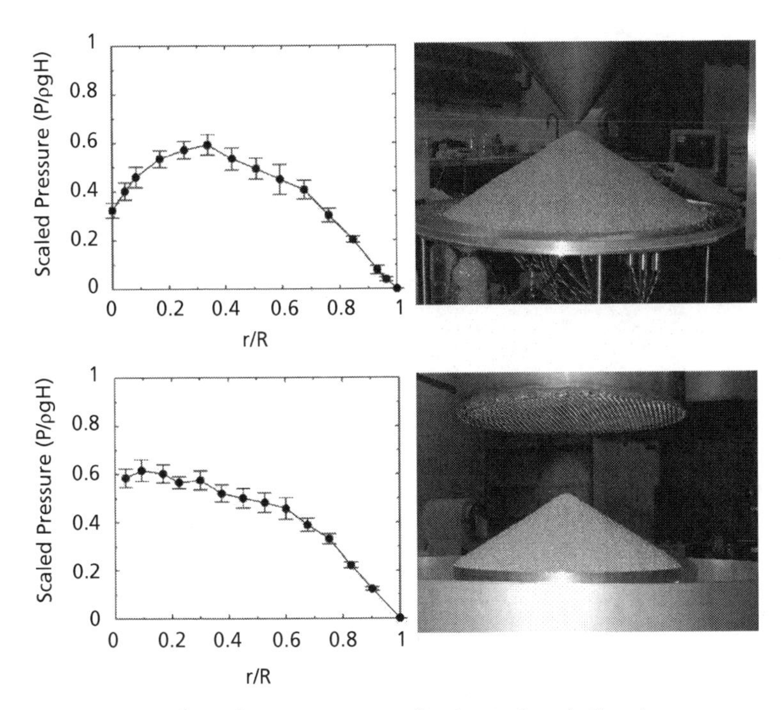

Figure 4.6 On the right are two outwardly identical sand piles: the upper one created from a single stream of sand raining onto the top point of the pile, the lower by a uniform 'rain' of sand from the sieve visible at the top of the picture. On the left are plotted the corresponding weights transferred downwards onto the supporting platform, as functions of the distance from the centre of the pile ($r/R = 0$) to the edge ($r/R = 1$). The pile formed from the single stream has a dip in weight under the central peak. Image courtesy of E. Clément, reproduced from Vanel et al. 1999. © 1999 by the American Physical Society.

laboratory of Tom Mullins.[10] The figure shows the greater compressive forces with correspondingly brighter illumination.

Immediately we can 'see' what is going inside the pile. Weight from above is not distributed uniformly, but channelled into 'force chains', splitting and dividing as they cascade downwards, and striking the base plate from above rather as forked lightning strikes the ground. Looking harder at these special paths, we notice two aspects of their behaviour that

[10] I. Zuriguel and T. Mullins, The role of particle shape on the stress distribution in a sandpile, *Proceedings of the Royal Society A* 2011, **464**, 99–116.

Figure 4.7 A visualisation of 'force chains' inside a two-dimensional sand pile. Image courtesy of T. Mullin and I. Zuriguel.

seem to be in tension. First, the chains are far from geometrical: they wiggle randomly on their way to 'grounding' the weight above them. They also pick up an apparently random value of force—there are weak chains, strong chains and everything in between (we see this from the varying brightness of chains in the figure). Second, and on closer inspection, there is after all some degree of order to the chains. Specifically they cluster around two preferred directions, one sloping towards the right and one to the left. This second observation is the clue to the dip in the force at the centre of the pile: the force chains constitute hidden 'bridges', buried internally within the pile. Just like real arched bridges, they do not collapse, because weight is transferred from the central sections to the extremities of the arch. The astonishing story with sand piles is that these hidden bridges of force chains arise spontaneously as the pile is laid down, without any artificial construction at all. *But they only do this when the sand piles are built up from a central stream.* When an identically shaped pile is formed from a uniform 'rain' of particles, the force chains are differently angled, and do not bridge over the centre of the pile strongly enough to create a dip in the force there.

A set of beautifully simple ideas developed by Sam Edwards in Cambridge and Michael Cates in Edinburgh,[11] together with their

[11] J.P. Wittmer et al., An explanation for the central stress minimum in sand piles, *Nature* 1996, **382**, 336–8.

collaborators, has indicated how this might be so. Think of the 'experience' of a single grain of sand landing on the pile in the two cases of uniform rain and central streaming. In the first the grain lands from above on a horizontal bed of material. It is supported from all sides and typically rests where it lands after some local minor readjustment of the grains forming its bed. The second story contrasts: a grain landing as part of the central cascade has a very different experience. Almost never does it stay where it lands (on the apex of the pile). Far more likely it finds no foothold and tumbles with its neighbours down the side slope of the pile. The slope itself is always in a special state. If at any point it becomes shallow, the slithering sand will come to a halt early, and so steepen it. If the slope becomes too steep, layers of sand from higher up will break away and avalanche down it, so making it shallower. The consequence is that, throughout the construction of the pile, its slopes are driven continually towards a critical angle at which landslides are always either close to starting or close to coming to a halt. How do the individual grains 'know' that this is the behaviour expected of them? This feature of nascent instability must somehow be encoded in their local packing structure: there is no other source of structural information local to the grains. This same structure in turn determines the directions in which local forces are transmitted to and from the grain—it can only push on its neighbouring grains at the points it touches them. So, and very subtly, the history of formation of the granular pile is encoded by the pattern of local contacts between the grains. These may describe a history either of raining from above or of tumbling down a slope and coming to a stop. But these different patterns of contact in turn determine the directions in which forces are transmitted through the grains after they become buried—structures that subsequently have to carry the weight above them as the pile grows. This is the simple idea of Edwards, Cates and company, and its legacy is the beautiful network of force chains that make internal bridges and become illuminated by Clément's and Mullin's experiments.

It is a delightful achievement to have understood the broad lines of the self-determining stresses in granular matter, but as always in science the doorways of our first glimpses of truth lead to great halls of unanswered questions. The delicate tracery of forces displays much more detail than anticipated by the simple assumptions of characteristic 'buried force' directions. More than that, there is no ability to anticipate the history of the avalanches large and small that build them, other than in a statistical sense. If this is true of a small sand pile constructed under laboratory conditions, how much more challenging is the problem of similar failures within the rocks miles below our feet, of which our knowledge is much less complete, and whose structure and consistency is much less homogeneous? But this is a beginning—these tentative theoretical and experimental explorations have begun to peer into phenomena that only a decade and a half ago were invisible to us. Asking statistical questions rather than exact ones has also proved again to be a wise move—we have made some progress rather than none at all. And nature has turned out to be rich in surprises once again.

Order Out of Chaos

Our journeys into some science motivated by storms above and earthquakes below have found two very different worlds in which predictability has vanished from even the most powerful methods our science can deploy. This seems at first sight to frustrate its goals, whether they are cast in the modern narratives of understanding, modelling, measuring and predicting or in the more ancient ones we have begun to unearth of wisdom, relationship and fruitfulness. Should we resign ourselves to pessimism at this point? Our examples suggest not.

The story of science is more subtle than that, turning not on demanding knowledge of everything, but on redefining its questions towards directions in which roads to understanding lie. Sometimes this requires a reframing of the problem in terms of the probabilistic and statistical, rather than the exact, nature of outcomes. In the case of the sand piles, such a statistical approach emerges as an understanding of the patterns of force chains, and how those patterns differ between piles of different history, rather than the prediction of all forces and structures in any one pile. In the chaotic pendulum it corresponds to knowledge of how likely it is to find the bob within any region of its domain, rather than a prediction of its exact course. Looking back at the time-lapse photograph of the double pendulum, we notice that, although the trajectories themselves wind about in the craziest of patterns, there is a higher 'density' of trajectory towards the bottom of the allowable space than the top—at this level it is possible to say, and to understand, *something* about the double pendulum rather than nothing at all.

Such a strategy seems better than an admission of failure, but even so still appears somewhat of a tactical withdrawal. More however is true, and we have caught sight already of ways in which new understanding might actually be opened up by grasping ignorance with both hands as part of our science (for this is what statistical questions do), rather than continually railing against it, or wishing to overcome it. Let us recall our 'stories of science' from Chapter 2 and look carefully once more at the peptide fibrils, and the remarkable Brownian motion which underpins their 'self-assembly'. We can now begin to recognise a connection between the chaotic molecular motion underlying that process, and the unpredictability of the phenomena of this chapter.

The concept of chaotic motion we learned from the double pendulum gives us new insight into the peptide problem. Looked at through the lens of molecular theory, the system of a peptide solution we were studying consists mostly of small molecules of solvent, sometimes just the boomerang-shaped water molecules—famously 'H_2O'. These attract each other enough to remain in close-packed proximity, but not enough to prevent a continuous seething, tumbling and dancing dynamics—the motion we call 'heat'. We can see straight away that if the two degrees of freedom represented in the double pendulum, together with the laws of motion, give rise to unpredictable trajectories, then even more so will this be true of such a system of billions of molecules. We can never have complete knowledge of how the dynamics of a solution unfolds at a molecular level of detail. Now imagine that we see a few larger, rod-like molecules (the peptides) added into the solution. They too cavort and diffuse, coupling their motion to that of the solvent molecules around them. There is a slight difference in attractive forces: those between the peptides, should they ever come into contact, are a little larger than those between peptide and solvent molecules. At the right temperature, these extra forces are just enough to overcome the buffeting from all around, and two peptide molecules will stay together for long enough on average to find a third. Then the little trio finds a fourth peptide to attract into the little huddle, just through the random side-stepping and tumbling induced by all the rolling water molecules. Something extraordinary is happening: a larger structure is emerging from a finer grained system, not *in spite* of the chaotic and random motion of that system but *because* of it.

Without the chaotic exploration of possibilities, the rare peptide molecules would never find each other, would never investigate all possible ways of aggregating so that the tape-like polymers emerge as the most likely assemblies. It is because of the random motion of all the fine degrees of freedom that the emergent, larger structures can assume the form they do. Even more is true when the numbers of molecules present becomes truly enormous, as is automatically the case for any amount of matter big enough to see. For now, although it is not possible to say exactly *where* the long assembled peptide tapes will be, we can be certain, within tiny margins of error, that there *will* be tapes somewhere, and even how their total lengths will be distributed.[12] Eventually a steady state of equilibrium is reached when as many tapes grow and join as are broken by the seething solvent. What can be stated as a broad probability for a few particles becomes more and more certain in the case of billions of billions. Out of the disorder emerges a new level of predictability.

This pattern of emergent structure from a substrate of chaos, embodied by examples like the peptide solution, is mirrored in the methods of the branch of science that has developed to explore it. Termed 'statistical mechanics', its conceptual and mathematical structure is rooted in the late nineteenth-century thinking of Boltzmann in Germany, Maxwell in Britain and Gibbs in America. As developed in the twentieth century it has become an astonishingly powerful tool, predicting not only qualitative phenomena, such as the appearance of self-assembled structures, but quantitative detail (such as the concentrations at which they form and the relative amounts of different structures). It is the conceptual world of temperature and the mysterious 'entropy', a term describing a formal way of defining 'disorder'.

There are much simpler examples all around us: the pressure of the air in the room in which you are reading this is (fortunately for you) a steady and predictable number. How should this be, given that what we call 'pressure' in a room is simply the aggregation of the effect of trillions of gas molecules colliding with its walls? There is, as we know, no way of calculating the motion of any of them for more than a fraction of its trajectory, so how could we ever predict a quantity that depends on all of them? Yet it is possible to predict the pressure with extreme accuracy if we know the density and temperature of the air you are breathing. It is possible, once again, not in spite of the random motion of molecules but because of it.

The calculation is very telling: the theory of statistical mechanics assumes in a strict sense that our ignorance of the underlying motion is

[12] It turns out that there are always many short ones, with long tapes becoming exponentially rarer.

total. At an early stage in any calculation within statistical mechanics, one needs to perform a very large sum, essentially a sum over all possible trajectories of the system, performed in a highly democratic way that shows no bias for one path or another. At a purely technical level, of course, these are rather harder 'sums' than the ones we do in school, and it takes a year or two more of mathematics classes to be able to do them, but conceptually they just add up the effects of all possible histories of all the molecules.

The idea is a beautiful one, but the outcome is wonderful: some of the patterns of behaviour of the system as a whole, and the quantities that describe them, emerge as *vastly* more likely than others, to the point of predictability for all intents and purposes. The exact pressure of a gas, the emergence of fibrillar structures, the height in the atmosphere at which clouds condense, the temperature at which ice forms, even the formation of the delicate membranes surrounding every living cell in the realm of biology . . . all this beauty and order becomes both possible and predictable because of the chaotic world underneath them.

Nature, it seems, operates on a statistical foundation, finding ways in which new levels of order can emerge from a chaotic substrate. So in asking these statistical questions ourselves, we find to our astonishment that, far from retreating from the project to understand the world, we are actually aligning ourselves much more closely with it. The emergence of prediction in our theories happens when we least expect it—when we have, in the face of a fine-grained world of chaos, given up on any hope of prediction. Even the structures and phenomena that we find most beautiful of all, those that make life itself possible, grow up from roots in a chaotic underworld. Were the chaos to cease, they would wither and collapse, frozen rigid and lifeless at the temperatures of intergalactic space.

This creative tension between the chaotic and the ordered lies within the foundations of science today, but it is a narrative theme of human culture that is as old as any. We saw it depicted in the ancient biblical creation narratives of the last chapter, building through the wisdom, poetic and prophetic literature. It is now time to return to those foundational narratives as they attain their climax in a text shot through with the storm, the flood and the earthquake, and our terrifying ignorance in the face of a cosmos apparently out of control. It is one of the greatest nature writings of the ancient world: the book of Job.

5

At the Summit: the Book of Job

We have dug enough now into the practice, history and experience of science to have perceived deeper layers within it than are suggested by its more superficial portrayals. Science runs far deeper, quirkier and at more fully human levels than we would think from stories of relentless discoveries, spectacular phenomena or the cool application of a fixed methodology. We know better than to swallow an inadequate narrative that portrays science as simply replacing an ancient world of myth and superstition with a modern one of fact and comprehension. Science, as we have framed it within a broader and older 'love of wisdom of natural things', does indeed call on a growing illumination of nature by experiment and imagination, creating understanding where there was none before and opening up the exploration of new phenomena. It maps, in increasing detail, the physical world onto patterns, often mathematical ones, in our own minds. Notably, the scope of science in both its experimental and theoretical explorations needs to capture the stochastic, the random and the chaotic as well as the regular, smooth and periodic. But science also emerges from an ancient longing, and from an older narrative of our complex relationship with the natural world. Its primary creative grammar is the question, rather than the answer. Its primary energy is imagination rather than fact. Its primary experience is more typically trial than triumph—the journey of understanding already travelled always appears to be a trivial distance compared with the mountain road ahead. But when science recognises beauty and structure it rejoices in a double reward: there is delight both in the new object of our gaze and in the wonder that our minds are able to understand it.

Scientists recognise all this—perhaps that is why when, as I have often suggested to my colleagues, they pick up and read through the closing chapters of the Old Testament wisdom book of Job, they later return with responses of astonishment and delight. Let us taste some of this beauty right away, from the point at which God finally speaks to Job (after 37 chapters of silence on his part) in chapter 38:4,[1]

> *Where were you when I founded the Earth?*
> *Tell me, if you have insight.*
> *Who fixed its dimensions? Surely you know!*
> *Who stretched the measuring cord across it?*
> *Into what were its bases sunk,*
> *or who set its capstone, when the stars of the morning rejoiced together,*
> *and all the sons of God shouted for joy?*
> . . .

We are by now familiar with this type of language; it is a beautiful development of the core creation narrative in Hebrew wisdom poetry, but now in the relentless urgency of the question form. The voice continues, initially with the familiar themes of boundaries and order, but with in the new grammar:

> *Where is the realm of the dwelling of light, and as for darkness, where is its place?*
> . . .

From the creative process of ordering, bounding and shaping it asks about the fundamental form of light, then sharpens its questions towards the phenomena of the atmosphere:

> *Have you entered the storehouses of the snow?*
> *Or have you seen the arsenals of the hail,*
> . . .
> *Where is the realm where heat is created, which the sirocco spreads across the Earth?*
> *Who cuts a channel for the torrent of rain, a path for the thunderbolt?*
> . . .

The voice then directs our gaze upwards to the stars in their constellations, to their motion, and to the laws that govern them:

> *Can you bind the cluster of the Pleiades, or loose Orion's belt?*
> *Can you bring out Mazzaroth in its season, or guide Aldebaran with its train?*
> *Do you determine the laws of the heaven?*
> *Can you establish its rule upon Earth?*
> . . .

[1] We take quotations of the text from the magisterial new translation and commentary by David Clines, *Job*, vol. 3. Bellingham, WA: Thomas Nelson, 2013.

The questing survey next sweeps over the animal kingdom:

> *Do you hunt prey for the lion, do you satisfy the appetite of its cubs,*
> *while they crouch in their dens, lie in their lairs in the thickets?*
> . . .
> *Do you know the birthing of the mountain goats, do you watch when the doe bears her kids?*
> *Can you count the months they must complete?*
> *Do you know the time when they give birth, when they crouch down to give birth to their young and deliver their offspring?*

It even addresses some of nature's puzzles and quirks of adaptation with humour:

> *The wings of the ostrich are beautiful, but are they the pinions of stork or falcon?*
> *For she leaves her eggs on the ground,*
> *and lets them warm in the dust, forgetting that a foot may crush them,*
> *and that a wild animal may trample them, . . .*

It wonders at the sheer power of a stallion,

> *Do you give the horse its might?*
> *Do you clothe its neck with a mane?*
> *Do you make it quiver like locusts?*
> *The majesty of its neighing is terrible.*
> . . .

and at the glory of flight in both its migratory navigational intelligence and mastery of the air:

> *Is it by your understanding that the hawk takes flight, and spreads its wings toward the south?*
> *Is it at your command that the eagle soars and makes its nest on high?*

From where does this voice come, a voice which resonates with question after question? The answer is itself a fascinating surprise. At the very start of this passage, known as 'the Lord's answer', we are told:

> *And Yahweh answered Job from the tempest,*

so situating the entire monologue within one of the wisdom tradition's great metaphors for chaos. Commentators have been quick to note that none of the animals appearing in the poem is domestic, nor are any of the cosmic powers of forces it asks about controlled by humans. This is an ancient recognition of the unpredictable aspects of the world—the whirlwind, the earthquake, the flood—that we looked at through a contemporary lens in the last chapter. It reminds us perhaps also of Steiner's horrified illumination of the 'inhuman otherness of matter', and his felt human need to be reconciled to it.

Even these short extracts from the longer poem give something of the impressive, cosmic sweep of this text, the grandeur of its scope, and the urgent, pressing tone with which it peers into the nooks and crannies of nature. In today's terms, we have in the Lord's answer to Job as good a foundational framing as any for the primary questions of the fields we now call cosmology, geology, meteorology, astronomy, zoology, ... Of course to use the text in that way is an unwarranted and anachronistic projection of our current taxonomies and programmes onto a quite different genre of literature and over a vast gulf of cultures and of centuries. However, if we are alert to the poetic forms which the narrative of early natural philosophy can adopt, then we can recognise in this extraordinary wisdom poem an ancient and questioning view into the world whose astute attention to detail and sensibility towards the tensions of humanity in confrontation with nature is unsurpassed. There are forces at play behind this text that lie at a depth and draw on an energy that still lies at the roots of the contested and developing relationship between the human and the inhuman. It can, in other words, speak into the old story of science.

Long recognised as a masterpiece of ancient literature, the book of Job has attracted and perplexed scholars in equal measures for centuries, and is still a vibrant field of study right up to the present day. David Clines of the University of Sheffield, to whom we owe the translation employed here, reproduced from his recent edition and commentary, calls the book of Job 'the most intense book theologically and intellectually of the Old Testament'. It belongs to a class of writing in the ancient world, indeed is its principal exemplar, concerning the 'fate of the innocent sufferer'. Many readers will be aware of the notions of 'the patience of Job' and 'Job's comforters' or 'Job's friends', which arise from the story.[2]

[2] Of many examples, the Akkadian work of a suffering poet, *Ludlul Bêl Nêmeqi*, the ancient Indian story of the righteous king *Harishchandra* rendered destitute then restored, and the Egyptian *Protests of the Eloquent Peasant* have all been contrasted with Job. Innocent suffering in the face of the gods was a regular theme in ancient Greek tragedy and philosophy from Homer to Herodotus who commented (7:46), '[suffering] demonstrates the gods' envy in that we have just begun to taste the sweetness of life'.

It is intriguing that, once again, ideas about nature are woven into a text that takes pain and suffering for its theme. However, although readers of the text have long recognised that the nature motif within Job is striking and important, it has not received as much comprehensive attention as the legal, moral and theological strands in the book.[3] This de-emphasising of nature might partly explain why the long passage of the Lord's answer, from which we have taken the extracts above, has had such a problematic history of reception and interpretation.

To rebalance a reading of these final chapters which is properly prepared for its nature imagery will need a brief survey of the whole book. We will not (the reader will be relieved to hear) be attempting here a comprehensive study of the text, though this is a very rewarding experience in the company of any of a number of excellent commentaries available today.[4] However, with the approach we are building of both the ancient narrative forms that tell of the search for understanding of the physical world and the contemporary stories of open-ended science, we ought to follow in more detail the way this astonishing book develops the human confrontation with the material universe. We will find it a rich mine of resources in our search to locate today's science in a much longer human narrative. At the very least it will take us to the highest pinnacle of all the ancient creation texts, from where we may glimpse more clearly the land we have travelled over so far, and plot out the territory we have yet to journey through. But first we need a rapid overview of the book's structure—a map within which we will later select our own path to follow.

The Book of Job: Theme, Story and Structure

The great theme within Job has been loosely termed 'the problem of suffering', but, as we will see, it actually runs far deeper than a lamentation in the face of pain and loss. It is the Bible's central offering on this theme, which furnishes a long philosophical tradition in all cultures. The problem of pain assumes religious forms: 'How can we reconcile the observation that extreme suffering occurs in the world with an omnipotent and loving God?', but it is worth recalling that secular versions also persist of the confrontation of experience with the assumption of a moral order in

[3] A notable exception are the chapters by K. Dell, N. Habel, D. Patrick and A. Sinnot in N.C. Habel and S. Wurst (eds) *The Earth Story in Wisdom Traditions*, Sheffield: Sheffield Academic Press 2001.

[4] The commentary by Clines contains an exhaustive list of precedents.

the workings of the universe. Bluntly, the complaint, as commonplace today as it was in the ancient world—'What have I done to deserve this?'—is Job's central question. The timelessness of the theme is deliberately suggested by its casting and language. One reason that the writing itself is hard to date is that it deliberately sets out to portray a very ancient time, the 'patriarchal period' before Moses when there were no priests among the Israelites, and before the term 'Yahweh' was commonly employed for 'God' (older usages such as 'Shaddai'—'Almighty'—are far more common in the book). Some commentators point out that the theme of unjust suffering as treated in Isaiah and Jeremiah, and associated with the captivity of Israel in Babylonia, resonates with that of Job, and in consequence place the composition in the fifth century BCE. However, there are no compelling textual reasons to do this, and much of the current material may well originate from earlier centuries.

The bulk of the text comprises monologues in Hebrew verse addressed between Job, his friends or 'comforters' and eventually from God himself. The poetry is framed by shorter narrative sections: an introduction, an epilogue and a short central comment. The structure is strongly reminiscent of theatre. The introductory prologue is, for example, employed as is common in staged plays, to give us onlookers information that the main dramatic characters are not party to. In the first opening narrative the curtains part to a conversation between God and a mysterious character called 'the Satan' (who should not be overlaid with the two-horned and pointy-tailed imagery of subsequent millennia). God points out the faithfulness of 'his servant Job' as an example of uncorrupted human righteousness, whereupon Satan rejoins that this is hardly surprising in the light of Job's immense prosperity and family happiness. He requests permission of God to bring suffering upon him, first in the loss of his family and possessions, then in the affliction of his own body with a terrible skin disease. Subsequent natural and human disasters bring these calamities about, so that the poetic discourses open with the afflicted Job sitting on an ash heap, despised and rejected by society. He is visited by three old friends, Eliphaz, Bildad and Zophar, and a younger acquaintance Elihu (whom we find out about only much later, probably as a result of the complex history of compilation of the canonical text). The friends sit with Job in silence and distress for 7 days before they begin to speak.

Then follow three textual cycles in which Job speaks out his complaint, followed by a speech from each friend in turn, each one answered by Job. There is a palpable crescendo of outrage as Job rails at God for giving him life yet then hedging it in with suffering. Increasingly he accuses his friends with indignation for their simplistic theologies and shallow arguments. Each fails in his own way to grasp the nettle of the moral problem, clinging instead to shallow but comforting moral 'certainties'. While Job points out the dissolution of the moral fabric in the world around him, and the absence of a moral order of retribution in the world, each in their own way insists that an easy law of rewards and penalties does indeed operate. Eliphaz claims that the suffering, while surely just, for reasons that Job is not admitting, will eventually pass. Bildad focuses on the sins of Job's children, which in his view must have brought about their demise, while Zophar accuses Job of demanding too easy a measured accounting of retribution and sin, unhelpfully pointing out that God has probably already been merciful to him in limiting the suffering he endures.

By now the reader of Job will have appreciated a slow transformation in, or a revealing of, the deeper nature of his complaint. For Job does not spend his breath demanding an end to his sufferings (this comes perhaps as a surprise if we assume that 'the problem of pain' is simply the main subject). His complaint rather is twofold: against the twisted *interpretation* and imputed *meaning* of his sufferings. The friends claim that the world embeds a 'law of retribution' (the evil are punished and the good rewarded), and deduce from it that Job has deserved his current fate. He turns the logic around: knowing his innocence he denies their accusations (6:15):

But my brothers have been as treacherous as the wadis.

(Note that here, as throughout the book, Job chooses a metaphor from chaotic nature to drive home his point—he is referring to stream beds that, though usually dry, fill suddenly and destructively after heavy rain.) From this he deduces his second complaint that there is no justice in God's world at all, but rather that its creator desires to trap and mock, even to torture, the humans he has made (16:12):

I was untroubled, but he shattered me;
He seized me by the neck and dashed me to pieces.

Any correction that Job might have merited from ignorance or oversight is insignificant, he claims, compared with the onslaught of punishment inflicted on him, which is at the very least out of all proportion. Job sees chaos, incommensurability and injustice everywhere.

As the debate develops, it moves relentlessly from pain-propelled philosophy to a legal dispute, where advocates are called for and formal positions articulated. Finally Job makes a remarkable move: he calls God himself into the theatre of the narrative, which has now clearly turned from ash heap into courthouse:

Will no one give me a hearing?
I have said my last word; now let Shaddai reply!
When my adversary has drafted his writ against me
I shall wear it on my shoulder . . .

At this point (chapter 32 in the book) a new character steps forward from the shadows, a younger man, Elihu, who has been an unnoticed witness to the events. He delivers a withering criticism to Job and the three older friends for their narrow-mindedness. He tries to weaken Job's complaint at one point by hinting that it is but a triviality before the greater purposes of God, and points out that suffering can act as a channel for warning. Yet he himself fails to escape in the end from reaffirmation of a shallow law of retributive justice.

Those are the events that precede the extraordinary moment at the start of chapter 38 when, contrary to all expectation (and as Clines surely self-referentially and endearingly puts it, can still create a frisson within those who 'have grown old with the Book of Job'), God does finally speak:

And Yahweh answered Job from the tempest, and said:
'Who is this who obscures the Design
by words without knowledge?
Gird up your loins like a man;
I will question you, and you shall answer me'.

There then follow the chapters of poetic questioning about the workings of the physical and animal world around and above Job with which we began. We briefly heard from the first of two discourses delivered by God to the now silenced Job (covering chapters 38 and 39). After a short and self-abasing acknowledgement from Job, a second discourse follows (chapters 40 and 41). It is in some ways even more remarkable, and certainly stranger, than the first, singing in exquisitely detailed praise the power of two monstrous creatures, the Behemoth and Leviathan (sometimes interpreted as the hippopotamus and the crocodile, but textually more splendid and terrible). These marvellous incarnations of strength and physical glory give the lie to any idea that biblical wisdom tradition identifies humankind exclusively as the pinnacle of the created order. On the contrary, this is a 'decentralising' text that places humans at the periphery of the world, looking on in wonder at its centrepieces. On Leviathan:

> *Beneath it are the sharpest of potsherds; it leaves a mark like a threshing sledge upon the mud.*
> *It makes the deep boil like a cauldron; it makes the water [bubble] like an ointment-pot.*
> *Behind, it leaves a shining wake; one would think the deep hoar-headed!*
> *Upon Earth there is not its like, a creature born to know no fear.*
> *All that are lofty fear it; it is king over all proud beasts.*

The book ends with a short epilogue, partly mirroring the prologue (but not reprising the heavenly scene with the 'Satan'). God reprimands Job's friends for not speaking what is right 'as my servant Job has'. So finally Job is justified. He prays for the friends that they might escape the just consequences of their folly (ironically playing for the first time in the entire book to a hint that there might be an operative justice of consequences in operation after all), and receives riches and family anew, living until 'an old man and full of days'.

The book of Job is a truly extraordinary story of surprises, beauty and replete with puzzles. It is, now that we have surveyed its structure in entirety, perhaps the last place in which we would expect to find 'the Lord's answer' in its present form of a long and questioning nature poem. Anyone attuned to the delight of discovery in the natural world treasures this text, but, situated as it is within the legal disputation of the suffering Job and his religiously self-righteous 'comforters', it has been problematic to readers. Does it really answer Job's two questions about his own innocence and meaninglessness of his suffering? Does the 'Lord' of the creation hymns correspond to the creator Yahweh of the Psalms, the Pentateuch and the Prophets? Does the text even belong to the rest of the book as originally conceived? Some scholars have found the Lord's answer to Job spiteful, a petulant put-down that misses the point and avoids the tough questions. Others, partly in sympathy with that interpretation, have suggested that the entire discourse has been 'glued on' to the earlier chapters at a later date and by a different author, pointing out that its ostensibly simple contrast of God's knowledge of nature to Job's ignorance provides no apparent satisfaction to his complaints. So, for example, Robertson[5] perceives that this 'God' fails utterly to answer Job, and finds him a 'charlatan deity'. Even those who take a very different view find the Lord's answer presenting an over-tidy view of the world. So Clines claims of it, 'There is no problem with the world. Yahweh does not attempt a justification for anything that happens in the world, and there is nothing that he needs to set right. The world is as he designed it.'

But are these interpretations justified? Increasingly scholars have recognised an underlying unity to the book that makes it hard to escape the tough questions around the Lord's answer by simply removing it to a distance or decoupling it from the preceding debates. Furthermore, its context of the painful aspects of humankind's relationship with the physical world is not, given the other wisdom literature we have already encountered, as unfitting as an unprepared reading might suggest. Looking at the text through the fresh lens of science today (even if that lens is at the wrong end of a telescope) resonates with the *difficulty* of doing science, even its painfulness, as well as its *wonder*—that is, how scientists respond at a first reading of it time and again. In particular, as we have noted before, it calls to mind the creative task that asks nature the right questions at the right time, questions that open up paths of understanding rather than lead to thickets of further confusion.

To bring these threads together we will take one more journey through the book of Job, but this time, rather than reading the small-scale map of its mountainous landscape, we will travel along the ground of a 'close reading', taking one path, albeit a not so well-trodden one, to the snowy peak where the 'voice from the whirlwind' speaks. For there is a track through the book that starts with the workings and structure of the natural world, and, while winding through the arguments of the disputations, never leaves it. This will be our 'nature trail' though Job.

Nature and Cosmos: a Nature Trail Through the Book of Job

Prologue

The natural world, and human stories about it, start appearing from the moment the prologue opens. Embedded in the backdrop to this opening

[5] D. Robertson, The book of Job: a literary study. *Soundings* 1973, **56**, 446–68.

scene are recurring images drawn from specific phenomena, ideas drawn from the natural order and from a selected group of creatures. These carry much of the illustrative weight that supports the negotiations between Job and his friends. So the Satan responds to God's boasting of Job's righteousness (1:8)

> *Have you not put a hedge about him and about his house and about all that is his on every side?*

calling on a two-edged biblical image. The 'thorn hedge' is the same as that imagined by the prophet Hosea (2:2) in keeping Israel from her idolatrous neighbours, but also brings to mind the protective wall God places around the Garden of Eden in Genesis following the banishment of Adam and Eve. It is an ambiguous boundary: is it protective (from enemies) or stifling (of a search for knowledge)? The ordering theme of the *boundary* between primitive chaos and created order that we noticed in other biblical creation songs appears, once again, 'in the beginning'. Satan's suggestion is that the boundary be removed.

The dire consequences are meted out through natural as well as human means (1:16):

> *The fire of God fell from heaven and burned the flocks and the servants. . . .*
>
> . . .
>
> . . . *suddenly a mighty wind came across the wilderness and struck the four corners of the house* . . .

Both the lightning and the desert wind (the sirocco, Palestine's hot wind from the east) are here no special miraculous interventions into the natural world. They do, however, carry superlative pointers ('of God' in the case of lightning and the hyperbolic striking of all four corners of a house in the case of the wind) that emphasise the enormity of their potential for violence. Both phenomena belong to the class of Job's recurrent natural metaphors for chaos.

The Prologue finishes with a shocking description of Job's physical state, covered in *grievous sores from the sole of his foot to the crown of his head*. The flood of natural disorder released by the removal of his protecting 'hedge' has finally curled its fingers around and into the flesh of his own body. From its outset the poetry points beyond a simple dualism of the human and non-human. On the contrary it presents a common natural physicality that blurs the distinction between the chaotic processes of storm and wind outside a body and the processes of disease and decay within it.

Job's first speech

Job chooses to begin his very first complaint not with a plea for suffering's end, nor with a head-on critique of (un)natural justice, but remarkably with a creation hymn, albeit a dark one, composed around not the first day of creation but the day of his own birth (3:4):

> *That day would it have become darkness! . . .*
> *that cloud has settled upon it, that eclipses had affrighted it!*
>
> . . .
>
> *Would that the curses of days had laid a spell on it, those skilled at rousing Leviathan!*
> *Would that the stars of its dawn had been darkened . . .*
> *and never seen the eyelids of the morning!*

Some commentators see in these verses a deliberate patterning, or rather a 'counter-patterning', against the creation story of Genesis 1. So Habel[6] points out not only the parallel themes of night and day, light and darkness and the creation of the sea monsters (Leviathan cf. Genesis 1:25), but also the comparison with the 'anti-creation' story of Jeremiah 4, which we examined in Chapter 3. Clines disagrees that Job's words are really meant to describe creation as a whole,[7] pointing out that the images are clustered locally around his own biography, not that of the world. But perhaps we are being asked here to continue travelling along the direction suggested by the Prologue—that control and chaos begin with the natural world, and that the human condition is intimately woven within it, not apart from it. This relationship is located in our mind through understanding, as much as in the fabric of our bodies. Job knows that days are turned to darkness not by magic but by eclipses and clouds and that Venus and Mercury (the 'stars of the morning'[8]) herald the dawn by the diurnal rotation of the skies. He even introduces us to great Leviathan 37 chapters before the great sea monster is invoked by God himself. In dark contrast to Yahweh's playfully triumphant romp, Job's Leviathan is all terror and curse.

[6] N.C. Habel, *The Book of Job*. London: SCM Press, 1985, p. 104.

[7] David Clines, *Job*, vol. 3. Bellingham, WA: Thomas Nelson, 2013.

[8] A point often lost in the analysis of the morning stars as a metaphor for hope is that they appear when the sky is still totally dark, and there is still *no other sign at all* of the dawning day. Their removal signals the loss of the last vestigial ground of hope.

The first cycle of speeches and replies (chapters 4–14)

The author of Job presents us with several subtle levels of irony in the words of the 'friends'. At the first level they are simply his companions, yet at the second become not friends but accusers. But, beneath that again, their brittle and narrow theology is sometimes used as a rhetorical pane of glass that if we look through it reveals, paradoxically, some of the deeper messages of the book. This device of operating on three levels begins with Eliphaz. Job's first friend picks up the powerful themes of nature that have already been, or which will be, presented. So (4:9ff.): The destroying wind of God:

> *By a breath of God they [the wicked] perish, By the wind of his fury they perish*

A divine voice from a whirlwind:

> *Then a wind swept past my face, a whirlwind made my body quiver.*
> *There stood a figure, unrecognizable; a form was before my eyes, and I heard a thunderous voice . . .*

The fruitful soil, and the flight of the eagle:

> *For it is not from the ground that affliction springs, not from the soil that suffering sprouts;*
> *It is a man who begets suffering for himself, and the sons of Pestilence [alt: as the eagles] fly high*

The natural wonders of God's creation:

> *He it is who works great deeds, past human reckoning, who performs wonders, beyond all numbering.*
> *He it is who sends rain upon the Earth, who pours down water upon the fields.*

And the eclipsed daylight

> *By daylight they [the wicked] meet with darkness, and at noonday grope as though it were night.*

Nature's glories to which the authentic divine voice will open Job's (and our) mind in all their freedom in chapter 38 are here presented, but then constricted, to serve Eliphaz's theology of retribution. For him, only the wicked receive the brittle fury of the whirlwind, and only the good the benefits of the transforming fertility of the rain. Readers from more northern climates need to be especially sensitive to the difference in connotation meant when writing about rain in the Middle East. This is not a seemingly endless late autumnal drizzling out of a slate-grey sky, but an essential recharging and transforming resource. It is a matter of survival or starvation whether rains come at the right time and in the right quantity. Eliphaz recruits such an essential phenomenon to promote his view of a world run on simple lines, so that creation is kind to those who deserve it, yet turns its back on those who do not. He presumes, as it has patently not been kind to Job, that his friend is in need of reprobation, and advises him to accept such discipline and look forward to a day when all is well once more. He chooses superlative terms in which to frame the rewards of Job's repentance (5:22):

> *At ruin and blight you will mock, and you will have no fear of the wild beasts.*
> *For you will be in covenant with the stones of the field, and the wild animals will be at peace with you.*

This is an extraordinary statement. The strength of the word 'covenant'[9] is such that some translators have shied from the notion of the implied blood treaty between humans and rocks, and weakened the sense. But we can hear a clear resonance with the strange 'distant vision' passages in Isaiah and Hosea we looked at in Chapter 3. In Hosea the strange covenant is with 'the beasts of the field' if not with other humans, but here the idea of a mutual relationship of understanding, reliance and permanence is explicitly extended to the inanimate stones, yet rooted in the covenant traditions of Israel's past. It reads as absurdly as a tale of a talking mountain, yet seems too powerful a notion to dismiss. Is the author of Job setting Eliphaz up by putting a foolish hyperbole on his lips? Is it an ironic and bitter reflection on the unrealistic dreaming of the tradition that surfaces in Hosea's vision? Or is it a paradoxical glimpse of the same deep story Hosea and Isaiah glimpsed, the 'third level' of irony into which the comforters sometimes speak?

Whichever way Job hears it, he twists Eliphaz's moralising on an ordered creation and builds on his own theme of the world out of joint. In chapter 6 he is the first to introduce us to the wild donkey:

> *Does the wild ass bray when he has found green grass?*

In other words, this wild ass (Job) is braying, like the animals often do, because, for no just reason, the world leaves them wanting, their needs

[9] The Hebrew *berith* is the same as used for the Abrahamic and Mosaic covenants of the Pentateuch.

unmet, their predicament one of pain. He turns the thrust of Eliphaz's invocation of life-giving water to rail against both his friends and God (6:15):

But my brothers have been as treacherous as the wadis.
They are like seasonal streams that overflow,
that are dark with ice, swollen with thawing snow;
but no sooner are they in spate than they dry up, in the heat they vanish away.

In one breath Job accuses God's universe of capricious disorder, rather than controlled provision, of holding out a promise of fruitfulness, then failing through chaotic unreliability to meet human need. He even manages a swipe at his friends by fashioning his picture of cruel inhuman nature into a simile for them. His nature language is used to portray both a world out of kilter with humans and, if it contains any moral law at all, a retributive repayment hugely out of proportion to human failing. So it is that we meet with a sea monster for the first time, a terror perhaps, in Job's eyes, great enough to merit the degree of restraint he seems to be experiencing at God's hand himself (7:12)

Am I Sea, am I the monster Tannin,
that you keep me under guard?

The next friend, Bildad, takes a new tack—of blaming Job's children rather than Job himself for his calamities—but his moral compass is as shallow as Eliphaz's. For him also, blame must be laid at someone's door, the only difference is whose. He, too, urges Job to consider lessons from creation, but, perhaps advisedly, tries drawing on the new arena of plant life for his parables. Papyrus withering in dry ground stands as a metaphor for a godless person, then in a confusing second image he begins to tell of a lush plant that only later is uprooted (8:16):

A lush plant is he in the Sun's warmth,
spreading its shoots in the garden.
Its roots twine about the heaps of stones,
it takes firm hold among the rocks.
But if it is once torn from its place,
that place disowns it with 'I never saw you'.

Bildad's point is surely to put Job under his accuser's spotlight by likening him to the plant once firmly rooted and grounded in the soil, but without deriving from it any guarantee of future security against one who comes to uproot. But his metaphor is somewhat subtle: the plant does not suffer from the rocky soil (in contrast to the famous 'Parable of the sower' in Luke 8) but rather adapts to its physical surroundings, enveloping the stones with its roots. We are reminded of Eliphaz's striking 'covenant with nature' by this picture of a 'covenant within nature'. The apparent opposites of hardened rock and living plants coexist within a reciprocal environmental niche—until, that is, both are overturned by external force. Once more a deeper voice speaking of something more permanent and significant seems to speak from the depths, yet perceived through the shallowness of the friends' superficial vision of nature.

Job responds (in 9:5–10) with another creation poem, but, unlike the human-centred versions of Eliphaz in chapter 5 or of Bildad in chapter 8, portrays God's power as equally destructive as constructive, without reference to human needs or actions:

He moves mountains though they do not know it; he overturns them in his wrath.
He shakes the Earth from its place and its pillars quiver.
He gives commands to the Sun so that it does not shine, and on the stars he sets a seal . . .
He alone stretches out the heavens, and tramples upon the sea-monster's back
He is the maker of the Bear and Orion, of the Pleiades and the circle of the southern stars.
He works great deeds, past human reckoning, he performs wonders, beyond all numbering.

Here is an ancient version of the juxtaposition we made in the last chapter—a presentation of the physical world in its chaotic manifestations of earthquakes and darkness, as well as in the ordered circulation of the stars, the named constellations and subduing of the ancient dangers of the sea. The topic of chaotic and ordered nature seems here to Job perfectly appropriate in the context of his search for vindication, even if it seems a strange choice of metaphor for us. For this creation hymn is launched with his new question, and an anticipation of its hopelessness in the light of divine wisdom and power:

How can a man be justified before God?
Should one wish to dispute with him, one could not answer him once in a thousand times.
He is wise and he is powerful; who ever argues with him and succeeded?

There is dismay in the 'inhuman otherness', not here of nature alone but of a God who, 'should he pass near me, I would not see him'. So Job makes the hopeless request for a mediator and courtroom in which his plea for justice might be heard. Imagining himself there delivering his complaint, he makes a remarkable further move. He extends his depiction of God's control over the world's physical structures, chaotic though they may be, down into those that make up his own body (10:9):

> *You moulded me like clay, do you remember? Now you turn me to mire again.*
> *Did you not pour me out like milk? Did you not curdle me like cheese?*
> *With skin and flesh you clothed me, with bone and sinews knit me together.*

The striking organic language of the curdling of cheese (a process of protein self-assembly related to the one we looked at in Chapter 2) and the ordered transformation of matter into living structures is the proof for Job that, since God's knowledge of him is complete at the level of every particle, he must actually know of his innocence. The chaotic dissolution of his flesh ('to mire') becomes a small-scale parallel to the unreasonable overturning of mountains and darkening of the Sun. The 'knitting together' idea echoes similar voices in the psalms (e.g. Psalms 139:13) that root humankind into all of natural creation, but Job then sets up, as he does time and again, a counter-narrative of the injustice of a creator who assembles living, self-conscious matter and then makes it suffer in uncontrolled decay.

The sunny and complacent Zophar is the last to speak in the cycle, and bases his dismissal of Job's complaint on an interesting second echo of Psalm 139, this time turning Job's quoted text back on himself. He entreats Job to recognise that the wisdom in creation's purposes lies beyond his grasp (11:7):

> *Can you uncover the mystery of God?*
> *Can you attain to the perfection of Shaddai's knowledge?*
> *It is higher than heaven – what can you do?*
> *It is deeper than Sheol – what can you know?*
> *Longer than the Earth is its measure,*
> *and broader than the sea.*

There is a fourfold dimensionality to Zophar's world, encompassing not only the attainable Earth, but the unattainable heavens and deeps. In an attempt to silence Job, Zophar recalls our travelling companion of the wild donkey (saying that its foal would more likely be born tame than an empty-headed man gain understanding), and inverts Job's 'darkness motif' by assuring him that his gloomiest future moments will shine as the noonday compared with his present gloom. As an attempt to silence the remonstrating Job, the reader is made to see that it fails conspicuously.

For Job completes the first cycle of speeches, not with silence but with a redoubled eloquence. For the first time he addresses the three friends together and at length (from 12:1 to 13:19), before redirecting his increasingly confident demands at God himself (to 14:22). His accusations directed at the friends are those of false wisdom, striking a man when he is down and that they are purveyors of worthless advice. He seriously envisages a legal confrontation with God for the first time, in spite of the hopeless terms in which he has previously spoken of such a court case. The second half of the discourse reads like a preparative account of charges, and focuses on the cruelty of God in recalling all avenues of hope from a dying man. To each of these twin purposes, Job invokes the nature imagery that has already appeared in the conversation, developing it with ironic detail. Clines and other scholars strongly advocate the view (for textual, grammatical and logical reasons) that the appeal to creation in 12:7–12 is a parody by Job of the simplistic views he has heard from his friends, so (12:7):

And yet [you say]:
Ask the cattle and they will teach you, the birds of the sky and they will tell you.
Or speak to the Earth, and it will instruct you, the fish of the sea, they will inform you.
Which among all these does not know that Yahweh's hand has done this?

The law of retributive justice and simple rewards is supposed to be so clearly woven into the fabric of the cosmos that the simplest of creatures can teach people about it. But by now we know that Job does not believe in a nature whose simple pendulum swings one way for the righteous and another for the wicked. The irony in this simplistic parody is not lost on us. It is of course doubly ironic that, come the Lord's final answer, it will indeed be nature that teaches Job the wisdom he needs to hear, but in a very different form from the nature lessons he has received so far. For now, Job reminds his friends just what forms divine 'wisdom and might' take in the natural world:

What he destroys will not be built, whom he imprisons will not be freed.
He holds back the waters, there is drought; he lets them loose, they overwhelm the Earth.

In contradistinction to an ordered cosmos supportive of human existence, Job emphasises the dark side of chaotic creation. Rather than the life-giving rains of Psalm 147 he points to the maliciously destructive power of waters in the wrong quantities at the wrong time.

His argument presses on to illustrate the hopelessness of existence, evidenced by the equally random fortunes of the wise and powerful among human rulers, then returns to two more natural phenomena to make the point. In contrast to the human condition of absolute loss with death:

But a man, when he dies, loses every power, he breathes his last, and where is he then?

Job summons up an imaginary tree, once felled yet budding anew with a tender shoot emerging from the stump (14:7):

For a tree there is hope that if it is cut down it will sprout again, that its fresh shoots will not fail.
Though its root grow old in the ground, and its stump begin to die in the dust, yet at the scent of water it may bud and put forth shoots like a plant new set.

Job's alienation from material creation is made even worse by comparison with such the 'covenant with the Earth' that trees seem to have by the property of their regeneration even after cutting to a stump (a horticultural practice known in ancient Transjordan in the case of figs, vines and pomegranates[10]).

Job's closing thrust at the end of the first cycle of speeches wraps up the argument for the illusory nature of hope by invoking a natural idea very rare in the ancient world (but not entirely absent—Lucretius refers to it in his *De Rerum Natura*[11]). He summons the properties of 'deep time' eroding through the ages the apparently (to humans) permanent structures of the Earth:

[10] David Clines, *Job*, vol. 3. Bellingham, WA: Thomas Nelson, 2013.

[11] There is a surviving record of an extended debate between Lucretius and Theophrastus on the finite or infinite history of the world in which Lucretius appeals to the erosion of mountains to support the necessity of a finite history, 'Do we not see rocks roll down, torn from high mountains, unable to endure the mighty force of a finite timespan? For they would not suddenly be torn away and fall if they had from infinite time past suffered without damage all the harsh treatment of ages.' Lucretius, *De Rerum Natura* vv. 315–317.

> *Yet as a mountain slips away and erodes, and a cliff is dislodged from its place,*
> *as water wears away stone and torrents scour the soil from the land – so you*
> *destroy man's hope.*

Commentators are divided on how this 'hopeless' aspect of nature should be read against the 'hopeful' phenomenon of the tree shoot. Are both or neither about the natural world? Is one rather than the other more significant of Job's currently nihilistic analysis of his predicament? But there is no reason to favour either usage—we know that the writer of Job is highly sensitive to natural phenomena, and to the human desire to ask questions of them, even to draw meaning from them. He is also deeply aware of the ambiguity of the natural world in relation to humankind, both in its direct physical channels (rain irrigates crops but also destroys them) and in the conceptual and reflective (the human mind can begin to grasp at an understanding of the world, even of the processes of time unimaginably longer than a human lifespan, but the vast ocean of nature's mysteries eludes us). Furthermore, when we look for meaning in the world, we witness as much chaos as order, as much tearing down as building up. At this stage in the argument Job can agree with his friends on one point: there is indeed a moral law woven into the physical fabric of the world that patterns the physical laws that operate there and equally within the material of the human body itself; however, it is not the well-ordered law of just rewards, but a chaotic law of ultimate decay and purposelessness.

The second cycle of speeches and replies (chapters 15–21)

The second cycle of speeches raises the rhetorical stakes. Although in disagreement and disapproval of Job during the first cycle, the friends' approach has been reconciliatory in tone, albeit superficial and brittle in content. Now they are losing patience themselves, their language becoming stronger and more condemning. One effect is that they dwell less on nature than in the first cycle: the metaphors become more loaded with human pictures of military action, hunting and prosperity. But if the theme of the natural world becomes subdominant in this second movement, it is never lost, and occasionally bursts out into the main thematic line. When plants, animals, rocks, atmospheric phenomena, light and darkness do reappear they are hurled across the debating floor with more force than in the first cycle. So Eliphaz calls on plant life as he did before—then recalling fruitfulness, but now to illustrate twisted or unnatural behaviour, as in the tree or vine of the unrighteous (15:32–33),

> *It will wither before its time, and his branches will not be green.*
> *He will be like a vine dropping its grapes while still unripe, like an olive tree shedding its blossom.*

Eliphaz's indignation at Job's failure to recognise the universal fate of the wicked emerges as accusation that they are, like nature, behaving unnaturally. He more than hints that Job's understanding has become similarly twisted. Job's reply also contains reference to the Earth; interestingly for the first time in the book in the form of a direct appeal. He first pleads (16:18) that it would, after Job's death, not let his case lie unheard

> *O Earth, cover not my blood, and let my outcry find no rest*

Then Job appeals to Earth's natural processes of decay to provide his body a final resting place. This second appeal marks the significant entry into his discourse of a new approach to a sort of reconciliation with disorder

> *If I have cried to the Pit, 'You are my father!'*
> *and to the worm, 'My mother!', 'My sister!'*
> *where then is my hope?*

The appeal to Earth and other structures within the cosmos exploits a practice found in ancient Egyptian and Hittite treaties, where 'Heaven and Earth' are called as witnesses to solemn agreements. It is a significant move at this point in the dialogue not only because it adds to its mounting seriousness but also because it builds another bridge between the parallel natural and legal worlds of the argument. We will see Zophar responding in kind when he next takes the floor, but first Bildad mounts an assault against the theme of decay, dissolution and destruction in Job's last two speeches. He flatly denies that cosmic disorder and decay constitute the eventual fate of the world (18:4):

> *You may tear yourself to pieces in your rage,*
> *but is the Earth to be unpeopled on your account?*
> *are the rocks to be dislodged?*

The idea of retributive legal order and cosmic structural order are, in Bildad's world-view, so inextricably entwined that he interprets Job's denial of moral justice to imply an affirmation of universal physical decay. He echoes a theme that we came across in late Isaiah in Chapter 3: that of the essential *peopling* of the Earth as an inherent element in the structure and ordering of creation.[12] Beginning to hint, as Eliphaz has begun to do, that 'the wicked' might include the unrepentant Job himself, he circles around the images of diseased flesh, destroying fire from above, the withering roots of a plant and perished farming land. No one is named as the target of his invective, but the reader is invited to hazard a guess (18:13),

> *By Disease his skin is devoured . . .*
> *Fire lodges in his tent, over his dwelling is scattered brimstone.*
> *Beneath his roots dry up, above his branches wither.*
> *His memory perishes from the farmlands . . .*

Bildad urges the need for a conceptual relationship with the world, but presents us with an unhealthy version of one, boxed in by formulaic falsehoods. The murky surface layer of simplistic shallowness has once more opened briefly to expose hints of a deeper layer of wisdom to the reader's gaze. The tiring Job knows this, but replies with repeated accusations of injustice by friends and God alike—Bildad may hint that Job's roots will dry up, but this is simply because

> *He [God] has torn my hope up by the roots.*

The slim hope of regeneration that Job appealed to before—in the picture of the felled tree stump—he sees now denied to him. Not even the minimally essential roots are left. Then in one of the most enigmatic verses of the book he reprises the theme of diseased flesh:

> *My bones hang from my skin and my flesh;*
> *I am left with only the skin of my teeth.*

It is puzzling in both form and content, for of course it is flesh that 'clings to' or is supported by bone, not the other way around, and teeth are one of the few exterior parts of the body without skin as a covering. Although some have assumed a corrupted text at this point, this is not a necessary conclusion, as Job is perfectly capable of the sort of hyperbolic irony expressed here: ideas that invert or exceed the normal course of nature. Like Hamlet's twisted metaphor prompted by the dark reflection that 'the time is out of joint',[13]

[12] Stronger still in Isaiah 45:18.

[13] *Hamlet* act 1, scene 5.

Job's world is so out of joint that his bones seem to rely on what is left of his skin to keep them together. Any scant remains of skin itself constitute furthermore an essentially non-existent portion—as much as is found on his teeth. It is a shocking portrayal of his diseased fabric. He would agree with Eliphaz that nature is behaving unnaturally, but not because of any immoral action on his part. His world is one of dark amorality not immorality, and that is due directly to the careless caprice of its maker.

Zophar continues to push the extremities of the nature image in his second-cycle contribution. There is nothing new he can add to the substance of his argument that the wicked man's moment of triumph is short-lived in the face of inevitable retribution. But far from taking leave of the natural world, he continues to invoke it, and now in superlative terms. Exploring the obverse of the 'floods and torrents' motif that Job has developed in its chaotic and harmful sense, he luridly depicts the beneficence lost to the wicked person (20:17):

He will enjoy no streams of oil,
torrents flowing with honey and cream

Zophar invokes and amplifies an old promise to Israel of a land 'flowing with milk and honey'[14] as a consequence of its people's obedience. But he cannot resist returning to the dark side of retribution, and as his speech reaches its climax he falls naturally into cosmic imagery once again

Heaven declares his guilt; Earth rises to denounce him.
A flood sweeps away his house, torrents on the day of God's wrath.

Here is the counter-appeal to 'Heaven and Earth' which matches Job's own raising of the stakes by his calling on Earth and Pit as family witnesses. We are not surprised when Job closes the second cycle by pointing out that neither the fecundity of their animal stock, nor the force of the whirlwind, nor the fabric of their flesh actually constitute a response of the Earth to the acts of the wicked (21:10,18,24):

Their bull sires without fail, their cow gives birth and does not lose her calf . . .
How often are they like straw before the wind, like chaff swept away by the storm?
His pails are full of milk, and the marrow is juicy in his bones.

[14] Exodus 3:8.

The animate, inanimate and human fields of the physical world all declare with Job that there is no correlation between morality and matter. The friends' case, supporting a retributive moral law, has already collapsed in the unjustified suffering that he, a righteous man, has undergone. It now fails again, for the second reason that the wicked receive no such consequence for their unrighteous acts. But Job's questions continue, and, as Clines has pointed out, not with the chief goal that the world adopt a different course, but more that he might understand the relationship of pain that is his current all-consuming experience of it.

The third cycle of speeches and the hymn to wisdom (chapters 21–31)

Perhaps you are familiar with Ravel's mesmerising and hot-blooded orchestral piece *Bolero*? For 20 minutes a weaving and insinuating theme twists and curls around the orchestra, turning back on itself, calling and answering to and fro as strings respond to woodwinds, and all the time, *all the time*, a relentless repeated *ostinato* rhythm on the side drum grows from an almost subliminal beginning to an overpowering insistence that possesses, then overwhelms, the attention. It is arguably the longest sustained *crescendo* in Western music, but there is one dreadful punctuating moment, marked by the entry of the brass, at which the whole swirling dance suddenly becomes, for the first time, *menacing*.

If *Bolero* serves as a simile for the twisting, repeating and rising arguments of the book of Job, then Eliphaz's entry at the start of the third cycle is like the entry of the heavy brass, the moment that makes explicit the latent menace within the cycling dialogues. Now at last all mere hints and insinuations that Job is actually to blame for his suffering, are replaced by clear and open accusation (22:5–11):

> *Is it not for your great wickedness, for your endless iniquities?*
> *You must have been taking pledges from your kinsfolk without cause,*
> *stripping them naked of their clothing . . .*
> *. . . that is why snares are all about you, and sudden terror affrights you,*
> *why darkness so that you cannot see, why a flood of waters covers you.*

This is of course pure speculation on Eliphaz's part, and from the outset the reader knows it to be wrong—that his own narrow theory of morality is forcing a perfectly unjust and unwarranted assumption. Job has indeed complained of darkness and flood—but that they randomly beset the just and the unjust alike. Eliphaz insists that they have fallen on Job (metaphorically—we know that these are not actually the physical forces that were unleashed on Job at the start of the narrative) for good reason—to punish Job for his wickedness.

There is a remarkable parallel at this point between the tone of the disputation and the material substance of the nature trail we are tracing through the book. At the very moment at which the personal accusations directed at Job become explicit, the nature imagery also climbs to a new dynamic level as we are treated (by Eliphaz) to our first explicit view of a Hebrew cosmology (22:12–14):

> *Is not God in the height of the heavens? Does he not look down on the topmost stars, high as they are?*
> *Yet you say, 'What does God know? Can he see through thick clouds to govern?*
> *Thick clouds veil him, and he cannot see as he goes his way on the vault of heaven!'*

The invocation of the starry heavens is the book of Job's equivalent of a full entry by the trombones. Its power is all the greater because cosmological structure is not common in Hebrew writing, but both here and in Isaiah (e.g. Isaiah 40:22) the 'vault of Heaven' makes a grand appearance as a giant inverted bowl over the Earth. God is positioned spatially high above a starry firmament, so is doubly exalted above the Earth—for the stars are unknowably high themselves. Eliphaz then invokes a second ancient tradition denoting the transcendence of the deity—that of his cloud wrapping. The great Mosaic theophany of Exodus[15] famously hides God that way, but this cloudy screen in Job is far more distant, hiding not only the deity from human eyes but, in a strange denial of omniscience, God's own sight of the world. A giant cloud-draped theatre on a cosmic scale becomes the stage for Job's arraignment. So it is that the brilliance of the book's irony also reaches new heights at this moment, for the reader sees Eliphaz's accusations miss their target before Job next even opens his mouth. The nub of Job's very argument is of course that God is perfectly well aware of the sufferings borne by the righteous within his creation, and that he does nothing about it. If God were unsighted by his own clouds then Job's complaint would be of non-omniscience, not of injustice and inactivity. Like a giant planetarium, the cosmos becomes the screen on which the hollowness and narrowness of Eliphaz's vision is projected for all to see.

[15] Exodus 35:5ff.

Just as Eliphaz has mapped out the dome of the heavens above in his escalation, so Job responds at equally grand perspective with a geographical map of the circle of the Earth below (23:8,9):

If I go to the east, he is not there; or to the west, I cannot see him.
In the north I seek him, but I see him not; I turn to the south, but I behold him not.

Neatly Job completes our first glimpse of the cosmos on the largest scale, and in contrast to Eliphaz, and as we might expect, uses it to reaffirm his innocence

For he knows what is my way, . . .
. . . my feet have kept to his path, I have followed his way without swerving

Earth's compass points become the frame for a journey of innocence rather than blame. Job insists that God does indeed see him; no cloud obscures his knowledge of Job's journey. Woven into the geographical picture so tightly that I have excised it for clarity from the short quote above is another allusion to metallurgy

Should he assay me, I should come forth pure as gold

but of course this belongs in an 'earthy' passage where the ground below is extracted and purified in the crucible. Its porous walls would absorb the molten base metals originally admixed with gold, leaving only the pure metal cupped within. Job insists that God's knowledge of him is of such chemical intimacy and that his righteousness is as pure. In the full orchestral *tutti* of the third cycle, this reads like a high trumpet line within the surrounding musical canopy of the bass.

As the high-pitched and offensive tactics of the third cycle get going, Job does not stop at defence, but now accuses God of failing to bring the wicked to justice when they steal the land of the defenceless, throwing them by force out of their homes. Our friend the wild donkey makes another appearance, invoked by Job as a simile for the fate of the homeless (24:5–8):

Like onagers of the steppe country, they go out to their work,
foraging for provisions, and the desert yields them food for their children.
They reap in a field that is not their own, and glean in the vineyards of the wicked.
They pass the night without clothing; they have no covering against the cold.
They are drenched by the mountain rain, and for lack of shelter take refuge
among the rocks.

Job's use of nature continues to be more subtle than that of the friends—he alone recognises the ambiguity of human and animal relationships with the world. In the most unlikely niches there is some sustenance, and although the mountains' cloud-tops hurl down the worst of the rain, their crags also provide refuge. The language of rocks probably refers to caves within mountain terrain rather than bare exposed faces, - there is a sense of an 'embracing' of the landscape that echoes the striking idea of a 'covenant with the stones' we remarked on in the first cycle. But none of this satisfies Job's inner desires for justice and explanation.

No more satisfying is the third-cycle declamation of Bildad, who stokes the rising temperature of the disputation. He ignores Job's real demands and instead shapes the cosmological backdrop of this movement into something rather fearful. Like the point at which the single side drum in *Bolero* seems to turn into an entire army on the march, he turns the skies into a military field of action (25:2,3)

> *A dreadful dominion is his; he imposed peace in the height of his heavens.*
> *Is there any number to his troops? On whom does his light not arise?*

Elsewhere Hebrew thought pictures the starry host as God's army—a protecting force—but here they are the ugly perpetrators of a totalitarian regime which reigns by fear. As a counterweight to the same leverage that he has generated from the third cycle's cosmology to elevate God to the status of tyrant, Bildad manages to reduce humankind to the level of invertebrates (25:5,6):

> *Behold, even the moon is not bright, and the stars are not clean in his sight.*
> *How much less a mortal, who is a maggot, a human, who is a worm?*

It is worth noting another important distinction between Hebrew and Greek cosmology that is exploited here. Far from dwelling in the superlunary medieval spheres of 'perfection' that belong to an Aristotelian cosmology, the stars here share the fallen imperfection of the world. This spatial homogeneity of category allows a degree of connection in Hebrew thought between the stellar and earthly domains that contrasts with most other ancient cosmologies (and which will arise again later in the Lord's answer), but here gives Bildad an excuse to exile humankind in value, as Eliphaz had just done in visibility, two giant leaps (as far from the moon as it is from God's heaven) from the deity.

The reign of fear imposed from on high extends right down to the shadowy realm under the Earth (26:5,6):[16]

> *The shades tremble in terror beneath the waters and those who live in them.*
> *Sheol is naked before him, . . .*

Bildad completes his tour of the terror-stricken universe by pulling back his viewpoint to give us a perspective of the whole construction (26:7–10)

> *He it is who stretched out the North over chaos, and suspended the Earth from nothing.*
> *He wrapped up the waters in his clouds, but the clouds did not burst under their weight.*
> *He covered the sight of his throne; he spread his cloud over it.*
> *He drew a circle upon the face of the waters, as the boundary between light and darkness.*

This is clearly another 'creation story': as much a description of the creative act as it is of the current and resultant structure of the universe. The 'North' here is a little puzzling, as it is clearly not simply the geographical direction but is more appropriate to the dynamics of construction of the heavenly fabric itself—perhaps stretched out from south to north. Astonishing suspension is everywhere—clouds hang, freighted with tons of water but without support; the heavenly vault is likewise without a central pillar, and the Earth itself at the centre hangs from nothing at all. Metaphorically we even catch a beautiful sight of the idea of order 'suspended' over the sea of chaos. But the beauty once glimpsed is shattered by the terror of it all. In a reference unique in the Bible to 'the pillars of heaven' we find fear even there (26:11)

> *The pillars of heaven trembled; they were aghast at his rebuke*

The third cycle of speeches is indeed a brutal one. It seems that the only way that Bildad can answer Job's searching complaints is to attempt to trump them with a charge of irrelevance. God, all-powerful, may rule with an iron fist if he so desires. There is no ethical complaint permitted against the one who defines all meaning, all ethical value in the first

[16] Here we take Clines' proposed re-ordering of the text to assign 26:1–14 to Bildad rather than to Job, in the face of strong evidence of corruption of order in the text from chapter 24 to 28.

place. Finally there is no human voice worth hearing in such a world: all voices are crushed into silence.

Composers have a hard job when the piece of music they are shaping reaches an ugly and deafening climax. Where do you go when the full forces of the orchestra have once joined their different themes together, taking the listeners to a moment of overwhelming power but also of terror? Perhaps this is one reason why the later editors of the text get confused at this point, why some versions assign no third-cycle speech to Zophar at all, and why Job's responses become snatched and stylised. It happens in music too: themes disperse, shattered into different voices, attempting to regroup, to find an answer to the experience they have just lived through (think of the aftermath to the thunderstorm in Beethoven's *Pastoral Symphony*, or hymn at the dawning of day at the close of Mussorgsky's *Night on a Bare Mountain*). The alternative, as adopted by Ravel in *Bolero*, is of course just to end the whole thing right there. Some commentators on Job have suggested exactly that of the original text. But that is not what happens in the version we have received from antiquity. More like Mussorgsky than Ravel, the disordered and frightening impasse is broken into by a new voice singing a new subject. Or perhaps on a closer hearing it is really an old theme, deeply buried but now resurfacing. Let us listen to the start of chapter 28, a beautiful and structurally quite new voice sometimes denoted the 'hymn to wisdom' (28:1–6):

> *Surely there is a mine for silver, and a place where gold is refined.*
> *Iron is taken from the soil, rock that will be poured out as copper.*
> *An end is put to darkness, and to the furthest bound they seek the ore in gloom and deep darkness.*
> *A foreign race cuts the shafts; forgotten by travellers, far away from humans they dangle and sway.*
> *That Earth from which food comes forth is underneath changed as if by fire.*
> *Its rocks are the source of lapis, with its flecks of gold.*

The scene is a mineshaft under the ground, and the voice is a miners' song. Foreign workers in the ancient Middle East were commonly employed in such dangerous occupations; here, we picture the mineworkers tunnelling and cutting deep underground. Roped to the subterranean rock face, we can just make them out swaying in the gloom. We also begin to see *with* them: a miner's gaze on the Earth from below-ground reveals a very different appearance from that above. The 'transformation as if by fire' is a remarkable insight into one of the processes by which minerals separate out, recombine and solidify in the rocks below ground. If we look hard in the dim candlelight we might catch a glimmer of gold. The discourse of earlier chapters touched in one or two places on smelting and refining metals from ores, but this song takes us much further back into the process of extracting the ores, and in an entirely new setting. It even begins to probe how the ores might have arrived there – the Earth itself has been a crucible for them. The underground world takes us completely by surprise—why did either an original author or a later editor suppose that the next step to take in the book was down a mineshaft? Reading on,

> *There is a path no bird of prey knows, unseen by the eye of falcons.*
> *The proud beasts have not trodden it, no lion has prowled it.*
> *The men set their hands against the flinty rock, and overturn mountains at their roots.*
> *They split open channels in the rocks, and their eye lights on any precious object.*
> *They explore the sources of rivers, bringing to light what has been hidden.*

We begin to recognise a tune that has been with us all along our nature trail through Job—it carries the theme of something especially human about the way we fashion our relationship to the physical world. It affects where we go (*There is a path*), what we see (*unseen by falcons*), what we understand (*bringing to light what has been hidden*) and what we do (*split open channels in the rock*). This extraordinary power to connect with nature seems so strongly worded that some readers have assumed that here the song is really talking about God, the creator himself, not humans at all. To take a specific example, 'overturning mountains at their roots' sounds like the exercise of divine power, but, if we bear in mind that the Hebrew word (*haphak*) translated as 'overturning' is just the same as that for 'changing' (as in 'changed by fire'), the metaphor directs us rather to admire the patient and knowledgeable art of mining seams through hard rock, exploring just those places that yield precious ores or stones. Perhaps the creation of pilings around mineshafts makes the writer think of the ploughing over of fields in agriculture. But even more significantly, once having gained access like no other creature (the gaze of both birds and mammals is restricted to the Earth's surface) to the underground world, only human eyes can *see* it from this new viewpoint. This is not purely impressionistic 'seeing'—the care with which eye, mind and hand are brought together in the description of the mine do not allow such a purely passive interpretation. It is a sight that asks questions, that directs further exploration, that wonders. No wonder such extraordinary human capacity has been confused with the divine.

Until now, the writer has kept to himself the primary subject of the hymn in chapter 28. Revealing it finally answers our question of what brought us underground in the first place. Now we are let into the secret, with the sounding of a new question (28:12):

> *But where is wisdom to be found? And where is the place of understanding?*
> *Humans do not know the way to it; it is not found in the land of the living.*
> *The ocean deep says, 'It is not in me', and the sea, 'Not with me'.*

We have been on a quest for wisdom and understanding all along. These are two different Hebrew words as they are in English. They mean two different things: the first a general idea of practical knowledge, the second a more intellectual and contemplative grasp. Carol Newsome in her commentary on Job points out that their juxtaposition can signify something more, 'the kind of understanding that would provide insight into the nature and meaning of the entire cosmos'.[17] And a timely search it is—taking us as it does out of the circling arguments of the disputation, but by no means irrelevant to them. Surely wisdom is what the court of appeal will need to resolve Job's demands and the friends' unsatisfactory replies.

Perhaps the idea—that the type of wisdom we will need is the same that searches the workings of the cosmos—does not now seem quite so strange after our journey through the circling arguments themselves, and their continual return to the natural world. But the poet teases his readers by misleading us—we are led to ask 'the wrong sort of question'—a naïve search for a 'place' where wisdom might be found. It is nowhere on land, nor in the oceans. Perhaps it is found in the treasures unearthed by those excavations deep in the mine? We follow the trail of the gold and precious stones from the mine to the marketplace, but find that if we ask there for wisdom (26:15):

> *It cannot be bought with refined gold, and its price cannot be weighed out in silver:*
> *It cannot be valued against gold of Ophir, against precious cornelian or lapis lazuli.*

[17] C. Newsom, The book of Job. In *The New Interpreter's Bible*, vol. 4 (eds L.E. Keck et al.). Nashville: Abingdon, 1996; quoted in David Clines, *Job*, vol. 3. Bellingham, WA: Thomas Nelson, 2013.

Gold and glass cannot equal it, and jewels of gold cannot be exchanged for it.
Coral and rock crystal are not worthy of mention beside it; a pouch of wisdom fetches more than rubies.
The olivine of Cush cannot compare with it, it cannot be valued against pure gold.

This is an indulgent and rich passage—Newsom draws our attention to the five different words used for gold alone. We find on the stalls in front of us precious things from both the deep Earth and the deep sea, but not what we are looking for. Has the poet led us on a wild goose chase?

Perhaps things are even worse—that the reckless tunnelling out of the rock for jewels and metals are, as in Ovid's account of the Four Ages of the World in his *Metamorphoses*, a terrible unleashing of danger?[18] The writer tells us that it is hidden from the eyes of all living things (humans, presumably, included), and that even the deeply buried land of death has only heard enigmatic whispers of it. So is the world simply a collection of 'the wrong places to look' for wisdom and understanding?

The conclusion of the hymn has driven different readers to opposing views. The end of the hymn 'draws back the curtain' once more in the search for wisdom (28:23):

But God understands the way to it; it is he who knows its place.
For he looked to the ends of the Earth, and beheld everything under the heavens,
So as to assign a weight to the wind, and determine the waters by measure,
when he made a decree for the rain and a path for the thunderbolt –
then he saw and appraised it, established it and fathomed it.
And he said to humankind,
Behold, wisdom is to fear the Lord, understanding is to shun evil.

The answer is for some commentators a deep disappointment, even a banality (e.g. Clines)—is it really true that, after all this exploration of nature from above and below in a search for wisdom, we find all that to be futile and are left only with a simplistic recourse to a pious 'fear of the Lord'? For an imaginary version of humanity without history, or rather without an intellectual future story, perhaps this is so. But the wisdom hymn by no means concludes that wisdom has nothing to do with the created world, for the *reason* that God knows where to find it is precisely

[18] Ovid *Metamorphoses*, Book I:136ff.

because he 'looked to the ends of the Earth, . . . , established it and fathomed it'.

It is, as for the underground miners, a very special sort of looking and fathoming—involving number (in an impressive leap of the imagination in which we assign a value to the force of the wind), physical law (in the channelled paths of rain and lightning) and formation (there is a blurring here between creating the world and looking at it once it is created). This is an extraordinary claim: that wisdom is to be found in participating in a deep understanding of the world, its structure and dynamics. It is banal and disappointing only if the very final injunction to humans is to be taken as an instruction to turn from numbering, weighing and participating in nature as God's exclusive domain, and restrict our thoughts and behaviours to the moral sphere alone. Is the writer really saying that that sort of 'fear of the Lord' is the end of all wisdom?

If we do conclude this then I am not sure that we are listening closely enough to the latent meaning of the word 'beginning' in the parallel, and well-known, opening to the book of Proverbs, 'The fear of the Lord is the beginning of wisdom' (Proverbs 1:7). Wisdom, in the writings we looked at in the last chapter, is not a state but a path (the writer of Job would employ what is becoming a favourite word: *derek*—'a way'). If there were no prospect of escape from the upwardly spiralling arguments of the book of Job so far, then we might make a different, more hopeless, interpretation of the human condition. But we know already that the momentum of the book will not remain circular, nor will God always remain speechless, but will extend to Job, and to the readers of the book, an extraordinary invitation to engage with him, and especially with his 'fathoming' of the universe. If the 'fear of the Lord' carries a higher meaning of engagement, following and exploration, rather than a simple and resigned moral obeisance, then the end of the hymn to wisdom is far from a banal journey's end. Instead it becomes a signpost to a mountain top view currently obscured from us, but in our nature trail through the book now only just around the next outcrop.

That final turn of the trail develops the idea that creation itself has potentially a voice, that it might become a participant in the future of Job's story. It is taken up in a final, reflective and sombre speech of Job himself, and in the words of the participant who has been silent until now—the younger companion Elihu.

The final speech of Job and the creation song of Elihu (chapters 31–37)

As we have already observed, the text of these final chapters before the Lord's answer is rather confused, and shows clear signs of rearrangement from its original sources. Although the current ordering of the text is unsatisfactory, there is no single alternative that stands out as superior among all others. What is clear is that the writer or arranger of the book introduces a new voice in this section: that of the younger companion Elihu. Clines makes a strong case that the hymn of wisdom constituted the original climax and close of Elihu's long speech. If that proposal is correct, we would then expect to encounter it immediately before the final appearance of the voice from the whirlwind, rather than, as currently placed, immediately following the rebarbative third cycle of speeches. But, as we saw, it performs a necessary role where it now stands, defusing the frustration and deadlock of the disputation. It is pointing us beyond the quest for justice, or even the explanation of injustice, to a quest for wisdom.

It is not made clear to the reader of the book of Job whether this is just a diversion or whether, by 'wisdom and understanding', the way will be found in the end to the problem of justice. Perhaps the exact attribution of who is saying what in this section is a secondary question to the content behind the voices. In our climb towards the pinnacle of the book, we seem to have hit a vertical precipice—the hymn to wisdom sends us in a different and indirect trail towards the peak, like a turn onto a contour path. It poses some new questions, and points towards a place to look for answers. That place is, surprisingly, the cosmos itself, already recruited time and again by Job to illustrate his argument that God's behaviour is arbitrary and morally ambivalent, and by the other friends in their attempts to prove the contrary. But now the characters in the tableau before us, as much as we the onlookers, begin to wonder if nature is quite as tame as this—simply a source of illustrations for our arguments.

Job himself employs a new voice in this section, less shrill and more reflective, as he begins to examine his own past. In a long enumeration of possible temptations and failings, he opens his heart to us, and, while always maintaining his innocence, betrays much more of his inner desires and values than he has when engaged in direct argument. He surveys the fields of his relations: with the people and rulers of the city, with the poor and needy, with women, with his own servants. Among this relational and reflective analysis of past temptation, he comes in time to his relation with the physical creation (31:24):

If I have made gold my trust, or said to fine gold, 'You are my confidence',
if I have rejoiced because my wealth was great, or because my hand had gained much.
if I have gazed with delight at the Sun when it shone, or the moon in splendour,
and my heart has been secretly enticed, and my mouth has kissed my hand –
this also would have been a punishable crime, for I would have been false to God above.

It is hard for us in a secular, (post-)modern society to recognise the personal depth of this admission of latent desire. It is far more than a brusque dismissal of any implied or actual accusation of nature worship. The language of 'gazing', 'delight', 'splendour', and even the imagined action of blowing a kiss of worshipful love to those celestial objects of adoration too distant to contact betray how close Job has actually come to this attitude to nature. The moon glimpsed briefly between sodium-orange clouds over a city sky as we emerge from a taxi is one thing; the majestic procession of a silvery orb hanging among the thousand stars of a velvety desert night quite another. Now add the traditions of Sun and Moon worship from neighbouring Egypt, Syria, Persia and Parthia, which would all have been well known to as educated a man as Job. The hymn to wisdom has even directed our gaze alongside that of the creator's into the divine fathoming of the world. There is clearly a task to do, understanding to win; we too easily condemn the ancient world when it guesses that the way to heal the gulf between the human and the non-human is for the former to worship the latter. Job feels the temptation keenly, and articulates it movingly, but he is equally sure that it is not the way forward for a servant of God.

Another approach to the human relationship with nature is offered in the final speech of all (in the canonical ordering), from Elihu. As do all the speakers before him, he paints his own interpretation of the panorama of nature: for him the cosmos is God's great teaching aid (36:22):

Behold, God is exalted in his power; who is a teacher like him?

. . .

For he draws up drops of water, and distils rain from the mist,

. . .

Who indeed can understand the spreading of the clouds, the thundering from his pavilion?

When he spreads his light over it, he exposes the roots of the sea.

. . .

For to the snow he says, 'Fall on the Earth', and to the downpour of rain, 'be strong'.
Once his voice is heard, he does not delay them.
He shuts everyone indoors, so that all may recognise that he is at work.

. . .

At the breath of God, ice forms, and the broad waters are frozen hard . . .
to fulfil his commands upon the face of the Earth.

Elihu's tale runs through the seasons in enumerating autumnal storms, winter ice, then later returns to summer heat. His attention to detail takes us on to new ground, for, whereas we have met the rain before, we have not imagined its cycle of formation from evaporation to mist and precipitation. Similarly thunder and lightning have crossed the skies of the disputation, but Elihu is the first to detail that lightning always arrives first, thunder following. We have seen the clouds (and even seen through them) but have not stopped to wonder at the phenomenon of their 'spreading' over the sky, multiplying as if from nothing. Remarkably, Elihu, while denying that we can know the answers to these mysteries of creation, actually hints at an understanding of clouds in his metaphor of distillation—it is not that clouds come from nowhere, but that the moisture already present in the air condenses into droplets large enough to scatter light, turning the air they occupy immediately from transparent to opaque.

Although denying human understanding of any of these phenomena, Elihu does one new thing—he gives creation a voice. It teaches things: that rain provides, that lightning obeys, that frost and clouds fulfil God's commands. He takes up the language of 'laying bare the hidden', as did the miners in chapter 28, but now, rather than digging at the 'roots of the mountains', the lightning itself illuminates the 'roots of the sea' (a strange idea even when we have already come by its source in the more familiar and Earth setting of the mine). The pedagogic metaphor extends even to the winter weather: it acts like a school bell—once everyone is indoors and unable to work they may instead contemplate the meaning of the storm raging about them.

Elihu's voice may seem to be making an attempt to respond to the call for wisdom rather than settling the question of natural justice. His appeal to Job to recognise that there is a message in nature more subtle than the one he is currently hearing likewise looks initially fresh, but in the end we see that his vision is as claustrophobic as that of his other companions. The cosmology is firmly anthropocentric—all is in relation to humankind. Worse, all is finally unfathomable. Elihu taunts Job with the futility of the very challenge he has set before him

Do you know how God arranges his works, how he makes lightning flash from the clouds?
Do you know the spreading of the clouds, the marvels of one who is perfect in knowledge?
You whose clothes are hot when the Earth lies still under the south wind,
will you, with him, hammer out the sky, hard as a metal mirror?

Elihu's cosmos turns hard and inhumanly back on Job, and although he is its centre and purpose, the picture of a metallic sky beyond either conceptual reach or artifice is the last image left to us by the disputation speeches.

The Lord's answer: reprise

We have climbed one of the many possible mountain trails through the landscape of the book of Job, ascending all the way, in our journey keeping continuously to the theme of nature. Looking back, it becomes possible to see the grandeur, and also perceive some large-scale structure, in the cosmological sweep of the text. There is a distinct order, for example, in the realms of creation explored predominantly in the three cycles of speeches:

First cycle: Earth, winds, waters, springs, stones, sea
Second cycle: plants, animals, vines, milk, honey
Third cycle: heavens, Moon, stars, Sheol, the far extremities of the world

This progression, as well as the increasing tension of the disputation itself, drives the narrative along while the universe of storms, lightning, rocks, plant and animal life, stars and Moon swirls around us. Centre stage is superficially the 'problem of suffering', but by chapter 37 it is laid bare to reveal a much more complex form than we met at first. Job is not demanding an end to his pain, at least not at first. His complaint is a double one. First, he rails that God is not, after all, in any form of control of the natural world, that creation contains in any sense a moral law. But second, and more fundamentally, he complains that there remains to humankind no possible way to understand the world.

He is out of covenant with nature, and needs, similarly to his double complaint, a twofold reconciliation with it. The first is a reordering of his physical environment of disease and poverty. The second is a comprehension of the apparent disorder of the world. No satisfaction comes from his friends—for them the universe enshrines a simple moral law, and chains humans to its defendant's stall. In such a world there is no escape from Job's puzzle, and no answers to his questions. The one thread that has not quite run out is the voice that started to sing, in chapter 28, a hymn to wisdom. But no one seems to know how to find such wisdom, how to hear from her, how to give Wisdom herself a voice.

The hymn of chapter 28 does give a clue to where wisdom might be found—and in doing so draws together the nature motifs employed almost continuously by Job and his friends since the Prologue. The move is, in retrospect, an astonishing one, for until then we have seen the natural world of the storm, wind, plants, rocks and stars as purely illustrative of the moral world of vindication and retribution. Creation is the backdrop to the foreground of the legal disputation. But we begin to suspect that nature needs to play a much more central role than that of illustrative scenery—for, according to the voice of the song, it is in his knowledge of nature's structure and workings that Yahweh alone knows the way to wisdom.

Looking back at the three cycles of dialogue with this new perspective, we see not only that they have taken us through ascending levels of the natural world itself, but that they have also introduced us to a series of different interpretations of what a human relationship with nature might look like. Getting this relationship right, we now understand from the cornerstone passage of the wisdom hymn, offers us the prospect of a route to the precious possession of wisdom, the quality most starkly lacking from the disputations. However, none of the perspectives offered from the clamouring voices within the book of Job has succeeded, for each in its own way has been found wanting. We have met each perspective along the way, not simply personified by one speaker, but emerging as distinct voices among, between and above the speakers. They are not so cleanly distinguishable that all readers will agree on their number and description, but I think that at least six have been presented along the way by the time Elihu finishes speaking.

First is the 'simple moral pendulum'—the story of nature as both anthropocentric and driven by a moral law of retribution. This is the central narrative of the first three of Job's friends. They each nuance it in their turn, for example by introducing a discounting filter that tempers cosmic punishment with mercy, but the underlying simplistic, deterministic (and ultimately barbaric) world-view is effectively unaltered throughout the conversation.

Second is the 'eternal mystery'—the story that speaks of God's exclusive understanding of nature's workings in ways that humans can never know. It is also anthropocentric, but through the centrality of a universal blindfold rather than by universal retribution. It appears among all of the friends' contributions at one point or another, especially in doxology, and in a different sense from Job's complaints that he is being 'kept in the dark'. It garners illustrative weight in the repetitive use of untamed animals, distant points of the compass and cosmological structures far removed from us in time and space.

Third is the 'book of nature' idea—the story in which nature constitutes a giant message board from its maker for those who have eyes to read it. Attaining its height in Elihu's speech, it also colours the way that the friends draw moral lessons from the rain, plants and heavens. Yet again, humans are central to this relationship just as are pupils in a classroom, but this classroom belongs in a kindergarten not a university. There are lessons to be learned, but this third narrative shares with the second the notion that the 'way' and 'place' of nature's components are hidden. The pupils who read nature's book in this way may be entertained, overawed, but they do not graduate.

Fourth is the story of the 'uncontrolled storm, flood and earthquake'. This is uniquely Job's interpretation of his relationship with nature, amplified by his personal anguish and exasperation. Through the lens of this perspective creation is chaotic rather than regulated, and is furthermore bound over to a crumbling decay. Humanity is swept up in the storm and flood, which God might have held at bay, but chooses not to. We might understand this perspective as the opposite of the first—there, order wins over chaos; here, the victory goes the other way.

A fifth possible relationship with creation is made explicit only once, by Job himself. It is the relationship of 'nature worship'. It is dismissed straight away, but not without giving away its allure. It presents modern readers with considerable challenges in our ability to empathise, but does serve to reignite a sense of wonder at the terrifying majesty of the cosmos. It also echoes an invitation extended to humankind to build some sort of relationship with nature while recognising that we can never be 'in control' of the world.

A sixth storyline is hinted at, but not spoken with clarity. It has something to do with the centrality of the created physical world over any claim by humanity to a pivotal place within it. It is the voice that locates wisdom within the knowledge of nature, and like the other narratives actually swims just below the surface of all the characters within the circle of disputation rather than finding a single proponent. It hints at a balance between order and chaos rather than a domination of either. It inspires bold ideas such as a covenant between humans and the stones, thinks through the provenance of rainclouds, observes the structure of the mountains from below, wonders at the weightless suspension of the Earth itself. But it is as yet unclear about where humans are located in its world-view, and agnostic about morality. It is the uncompleted storyline, and its 'loose ends' have become the inflamed nerve endings of Job's physical and mental anguish in a cacophony of voices that lead nowhere.

The book of Job reaches for the second time a point at which all its characters have fallen silent, but without resolution. The choice before writer, compiler and reader is again that between conclusion and the introduction of another, greater voice. We know already that there is only one voice, of course, that remains to speak.

This is the point at which we first entered the book of Job, but now I wonder if we feel more deeply the 'frisson within' as Yahweh finally speaks, at the very point where everyone else has finally fallen silent. He immediately announces what this new speech will finally do in the compact opening question—Job has '*obscured the Design*' by his ignorance, *yet* Yahweh is providing the answer Job has long demanded. The new voice signals that Job (and we the readers) should be very wary of interpreting what is about to come as a 'put-down' (in spite of the history of interpretation that takes exactly that line), for the invitation to 'gird up your loins' is spoken, shockingly, to a legal adversary of equal standing, not an inferior.

Job's pain and unsatisfied desire for vindication has led him inexorably to the thirst for wisdom, and to look for it in the unresolved 'inhuman otherness of matter'. He has explored several fruitless paths towards reconciliation with the nature that has his very flesh gripped in the claws of its apparent chaos and decay. Far from an irrelevant distraction then, the Lord's answer begins to fill out and extend the sixth and incomplete narrative. No diversionary tactic from Job's insistence on justice, Yahweh insists that to demand an answer to that question. while keeping the whole physical creation at bay, is a fruitless project. He plucks Job up from his ash heap and takes him, and us, on a whirlwind tour of creation from its beginning to the present, from the depths to the heights and from its grandest displays into its inner workings. We are reminded of the hymn to wisdom:

But God understands the way to it; it is he who knows its place.
For he looked to the ends of the Earth, and beheld everything under the heavens,

For in the poetry of his own speech we are now beginning to look with him, to align our gaze with his. Links with, and responses to, chapter 28 are suggested throughout the Lord's answer. The tone of the poem is hotly debated, but if we take together the threads of the explicit request to listen and answer, the poetic question form of the speech, and especially the momentum of ideas within the book as a whole, the voice becomes persuasively *invitational*. The 'grand tour' of recapitulation of not only the nooks and crannies of creation familiar from the earlier disputation but also the uncharted places special to the speech alone means that we, with Job, look further into creation and ask deeper questions of it than ever before. We begin to answer the call to perceive nature in a new way.

The change of perspective is immediately shocking. Job has accused God of ineffective control of the powers of chaos, but when we revisit the creative act of bounding the swirling oceans (38:8):

Who shut in the sea with doors?—when it broke forth from the womb,
when I made clouds its garment, and thick darkness its swaddling band,

we find that they are wrapped in infant's clothing as easily as a newborn child. Not content with simply recalling the phenomenon of the dawn, we expose its structure (38:12):

Since your days began, have you called up the morning, and assigned the dawn
its place,
so as to seize the Earth by its fringes
so that the Dog-stars are shaken loose?
It is transformed like clay under a seal and all becomes tinted like a garment,
as the light of the Dog-stars fades, and the Navigator's Line breaks up.

The process of dawn is, like creation itself, an outworking of the physical order with interlocking components. As light intensifies so the visual flatness of the Earth in twilight takes on the fully three-dimensional shape of hills and valleys. As sunlight increases so starlight is obscured and the constellations fade.[19]

The poetry reinforces the idea of an invited journey by use of its question form, unpicking as it does the idea from the hymn to wisdom of 'beholding everything under the heavens', so (38:16)

Have you journeyed down to the springs of the sea, or walked about in the remotest parts of the abyss?
Have the gates of death been revealed to you, or have you seen the door-keepers of death's darkness?
Have you gazed onthe furthest expanse of the underworld?
Say if you know its extent!

takes us, with Job, in our imagination to the very places that were searched out in the quest for wisdom. We recall that in that hymn the depths of the sea and underworld alike admitted that they had heard rumours of it but could not contain it themselves, but now the questions continually keep open the possibility of finding it. The voice is far from simply making a point by taunting Job with his now-exposed ignorance; it is showing him other ways of thinking about creation than in relation to his human predicament. It is beginning to invite him to think about it though the eye of its creator.

The tension between chaos and order, assumed in the earlier chapters as mutually exclusive and in combat with one another, is not ignored, but begins to be subtly resolved by this change of perspective in the Lord's answer. It affirms that forces of flood and lightning have their own special form of dynamics, in which they are neither predictably directed nor entirely uncontrolled, but are rather 'channelled' within a governance of freedom (38:25):

Who cuts a channel for the torrent of rain, a path for the thunderbolt,
to bring rain on a land uninhabited, on the unpeopled desert,
to satisfy a waste and desolate land, making the thirsty ground sprout with grass?

The surprise is twofold—first that chaotic energy is not denied but channelled, second that the resultant fruitfulness of the Earth has nothing to do with humankind at all. Job journeys with Yahweh to water and

[19] We follow Clines' suggestion that the common translation of 'wicked' for 'Dog Stars' arises from a slight scribal copying error in the Hebrew, supported by the absence of any moral referent in the entire speech.

refresh unpeopled lands where no argument about anthropocentric justice can even arise. In almost every stanza, the Lord's answer draws on natural imagery or phenomena already adopted by other speakers, but transports Job to his radically new viewpoint. The fascinating astronomical passage in chapter 38:31–33 that we met at our first encounter bears another look in this light:

> *Can you bind the cluster of the Pleiades, or loose Orion's belt?*
> *Can you bring out Mazzaroth in its season, or guide Aldebaran with its train?*
> *Do you determine the laws of the heaven?*
> *Can you establish its rule upon Earth?*

As we have already noted, this ancient Hebrew cosmos contains a closer connection between heavenly and earthly realms than the Aristotelian version with its essentially distinct division between the perfect and the imperfect at the Moon's orbit. Beyond the notion that physical law may apply equally to the stars as to the Earth, we should however probably not understand here the 'laws' and 'rule' in the Newtonian sense of universal physics (famously the idea that the same gravitational force accounts for the motion of a falling apple and the orbiting Moon). The connection is more likely the thought that the appearance of seasonal constellations in the heavens heralds the coming of annual rainy and dry periods on the Earth. The essential theme, though, is the idea of guidance rather than control, of a natural order that contains within itself openness, rather than a rigid predictability, and emergent order rather than an imposed one. It connects Heaven and Earth into one created system, with humans at the same time special because they are invited to participate in the wisdom of understanding it, but in no sense central or preferred. Again there is an invitation to think about how that order might come into being.

One by one, the forces of nature invoked by the disputants throughout the circling debates are summoned once again, but framed in a new way. The first five 'storylines for nature' we have heard on different lips are passed aside by the questioning, which instead urges its listeners to think about another way of understanding. So the challenge (38:34,35)

> *Can you lift your voice to the clouds, and make a flood of waters answer you?*
> *Can you send lightning bolts on their way, and have them report to you,*
> *'Ready!'?*

does not, as some have assumed, claim that God can actually command the clouds, the floodwaters and the lightning in a strict control of their every swirl and dart. It merely asks Job whether *he* can, urging him to think through a world that runs on a command-and-control economy. While laying divine claim to structuring the cosmos and creating boundaries for it, the perpetual question probes all our assumptions. Nature has its freedom, its 'way' and its contained and creative chaos. Just as it is the 'wrong sort of question' today to demand a predictive and ordered mathematical formula for the evolution of a chaotic dynamical system, so a world of obedient and ordered clouds was for Job the 'wrong sort of cosmos' to hope for.

If, rather, the notion of 'contained freedom' lies implicit in the accounts of inanimate behaviour, it becomes explicit as the enquiry passes on to the animal world. It revisits our wild donkey, by now surely beloved to writer and reader alike (39:5):

Who let the wild ass go free? Who loosed the onager's bonds,
to whom I gave the steppe as its home, the saltings as its dwelling?
It laughs at the tumult of the city, and hears no shouts from a donkey-driver.
It ranges over the hills for pasture, searching for anything green.

The wild ass symbolises here, as widely in ancient Middle Eastern literature, a counterweight to civilisation, the unruly world outside the city walls and urban legislation. So the point is not that God 'let the wild ass free' but that no one ever did: untamed from the beginning, it has always explored the open steppe and hilly ranges as its home. No rules dictate its gallop or direction any more than lightning or earthquake follow predictable timetables.

The climax of the total decentralisation of Job from his universe within the Lord's answer is the celebration of the two giant creatures Behemoth and Leviathan. As we have already met the latter, let us now hear the creator of the Behemoth praising its wonder (42:15):

Consider now Behemoth, which I made as I made you; it feeds on grass like an ox.
What strength it has in its loins, what vigour in the muscles of its belly!
It stiffens its tail like a cedar; the sinews of its thighs are intertwined.
Its bones are tubes of bronze, its frame like bars of iron.

We are reminded again of the 'interior' perception of a maker (or remaker) of a world. Like the miners who were 'overturning mountains at their roots' and so were able to see from their mineshaft the inner structure of the rock, so the maker of Behemoth is able to celebrate the detailed intertwining of its muscles that, in the balance of their tensions, permits the raising of a tree-like tail. His eyes even temper the structure and rigidity of the animal's bones. Once more the tone has changed: the tour of nature is still deeply perceptive, and still invitational ('Consider . . .') but it has now dropped the polemic tone of the 'where were you . . . ?' and the 'do you know . . . ?'. Job, and humanity with him, is now so decentralised that the voice is left in pure celebration. It is a resplendent climax of a journey to the mountain top of pain, questioning and glory.

One thundering question remains for all readers of the book: does Job receive an adequate answer to his two complaints in the Lord's answer? Most commentators suggest that they are bypassed; Yahweh's speech may be profound, beautiful and transformative, but it does not address Job's questions. By some measure this must be true, for we know by now that Job was asking in some sense the wrong questions. He has at least framed his questions in a way that distorted and pre-judged possible answers. But now we have seen that the question of God's justice in his management of creation as a whole is woven into Job's disputations, we recognise that it is not bypassing the question for the Lord's answer to take this thread and expand it into the glorious quest into nature's workings with which the book finishes (or nearly finishes). With trepidation, and against the weight of opinion, I am therefore suggesting that the 'Lord's answer' *is* an answer to Job's complaint—possibly the only adequate answer. There are five lines of argument that point this way.

First, it tackles head-on the accusation that creation is out of control by suggesting ways of thinking through what Job's (and his friends') idea of 'control' might mean. The deterministic and predictable response of a cosmos that metes out retribution on the unjust is not a living universe but a dead one. The axis of control and chaos is subverted by the revelation of a third path of constrained freedom in which true exploration of possibility, of life, really lies.

Second, it does, against all expectation, achieve what has always been Job's aim, to be reconciled to his state of physical pain and mental outrage. We are not privy to Job's inner response to the Lord's answer, we do not know the steps that lead him to aver at the end (42:5)

I have heard you with my ears, and my eyes have now seen you.
So I submit,and I accept consolation for my dust and ashes.

But we do know that he has been led towards a radically new perspective, one that in one way totally decentralises humanity from any claim to primacy within creation, yet in another affirms the human possibility to perceive and know creation with an insight that is at least an image of the divine one. By implication, the experience of pain cannot be divorced from the confrontation of human hope and a universe of freedom.[20] Its place within the book as a whole, and especially its resonances with chapter 28, suggest that this new perspective is foundational to the acquisition of wisdom. We are left to conclude that this reconstruction by combination of two moves in such extreme tension with each other releases the healing that Job so desperately needed.

Third, it is participative and invitational: the final voice asks the great questions about nature not purely to rouse into self-awareness Job's, or our own, lack of understanding, but as an invitation towards transforming it in encounter with wisdom. The possibility of a new relationship with the physical world is laid before Job that leaves behind the irresponsibly polarised positions to which he and his friends have been clinging. It recognises that although ignorance is always an aspect of the human predicament, and in particular of its confrontation with the physical world, it is not a static one. There is a history of knowledge with a past, a present and a future. Job and those who read about him are invited to assume a role within that history.

Fourth, it speaks of the fundamental significance and importance of the physical structure and workings of nature. They are not sideshow or an optional hobby for the socially challenged. Our relation of perception, knowledge and understanding is at the centre of our humanity. It may well be at the centre of our experience of pain—the failure to acquire wisdom in working with nature hurts both the world and ourselves. There are healthy and unhealthy ways of living the relationship between the human and the non-human, but in all cases the choice and its actualisation matter. They guide the development of our relationship with the natural world, which begins with 'seeing it in the right way', asking the right questions, exploring beneath the surface of things and participating in nature's history. The book of Job even talks of the direction of human relationship with creation in terms of a covenant. It

[20] Here I find myself disagreeing with Clines, who finds within the Lord's answer a view that 'there is nothing wrong with the world'. David J. Clines, *Word Biblical Commentary*, vol. 18C. Nashville: Thomas Nelson, 2012.

offers us the loftiest of all biblical perspectives on the programme we have called the 'love of wisdom of natural things'.

Fifth, the Lord's answer is 'eschatological'—its message is one that announces and urges the possibility of a future in which this vital relationship, now broken, becomes healed, not just for Job but for the species from which he comes. Job may not know the answers to Yahweh's questions, but one day he or his descendants might well do.[21] This is admittedly not the interpretation one comes to at first sight. However, it is a natural continuation of thought in the context of the nature theme we have carefully tracked within its course through the book as a whole. Especially in the light of the explicitly future-oriented search for wisdom in chapter 28, and in the wider Old Testament vision of a healed world, it becomes at the very least a plausible reading. Even the theatrical structure of the book of Job points at 'unfinished business', for although we return at the end to Job's reconstituted family, we never find an epilogue to match the prologue scene of the heavenly court. If such an eschatological interpretation holds, then Job must become an integral text for our purpose to unearth the ancient human wisdoms that have wound their course towards the community of scientific practice that we see today, for this is the very community that currently affirms that 'nature matters', that knowledge has a history, that it is participative, but that it is also risky.

A further strong line of evidence that the story of Job does not finish with the book of Job lies within the pages of the Bible itself. We have seen Job draw on other lines of thought within biblical creation stories and wisdom literature as it climbs to the mountain peak from which all the universe lies before its readers. In the same way we can journey on down the other side of the mountain and trace through the subsequent development of tradition how the agenda of reconciliation between the human and non-human evolves. The Bible itself embodies such a story of people, place and covenant. It is a story that continues, with the radically new twist of the incarnation, into the New Testament. Along that trail lies the next chapter of our story.

[21] This suggestion has been made at least once before, in the brief publication by David Wolfers, Science in the book of Job, *Jewish Bible Quarterly* 1990, **19**, 18–21. His conclusion runs, 'The majority of these questions [those of scientific purview] are to be found in the Lord's first speech to Job, and there is little doubt that their primary purpose is to expose the abysmal ignorance of mankind of all theoretical aspects of Creation. Is it possible to detect a faint hint, that it might be well for man to set about attempting to remedy this ignorance?'

6

Creation and Reconciliation: the New Testament Creation Narratives

From one point of view, the two parts of the Bible, the Old and New Testaments, are strange companions. The Old is a faith tradition's library of history, philosophy, poetry and genres less familiar to us today: wisdom, prophecy and apocalyptic. It draws on writers and editors over a period of at least half a millennium. The New Testament is much more hurriedly compiled: at most, half a century's writing. It responds to, and records events that led to, a seismic shift in the religious and political communities of the Mediterranean.

The New Testament contains four accounts of the life of Jesus, but each carrying far greater weight of meaning and implication for the community that reads them than would straightforward biography—they constitute the genre of 'gospel'. Then, after Luke's rapid history of the very early church known as 'The Acts of the Apostles', the rest of the New Testament is mostly a collection of letters ('epistles') by the leaders of the new 'Way', as Christianity was first called, to churches in the lands we now call Italy, Greece and Turkey. Some of these are the earliest of the New Testament documents, and, although positioned after the gospels in our Bibles, precede them by about a generation.

The period of writing and the cultural setting is important here also, for, by the time of the first century AD, we have moved from an Israel defined in relation to its neighbours to the south (Egypt) and east (Assyria, Babylon) to the advent of power from the west (Greek and finally Rome). Greek learning was now dominant across the Roman empire, and even those groups most thoroughly trained in Jewish thought such as the Pharisees (St Paul is an example) would have been familiar with Stoic and Epicurean philosophy. The engagement of Jewish and Greek thought is one of the most fascinating aspects of the New Testament writings.

Yet in spite of all these differences, for those second-century church leaders and theologians who debated the canonical content of the Bible, the Old and New Testaments belonged, quite literally, stitched together in their new 'codices'—the first paged books. They were for the early church part of one single story—the writers of the gospels and epistles were clearly convinced that the promises and hopes contained in the stories and songs of Israel were answered, and new challenges launched, in the life, works, death and resurrection of Jesus of Nazareth. The first chapter may have closed, but a second, unexpected one, opened. As the New Testament scholar N.T. Wright, among others, has pointed out,[1] these two 'chapters' are, from one narrative perspective, in strong continuity (the Old Testament hope that Israel's God would one day return to vindicate his persecuted people as Israel's king was indeed fulfilled in Jesus), and at the same time in radical discontinuity (the way in which this was to happen was through a radical transformation of kingship into servanthood, vindication achieved not in dominance but through suffering).

Such an interpretation of what the New Testament texts, and the communities that produced them, were trying to do should alert us to look out for this pattern of continuity and transformation in any of the ways in which the Old Testament speaks of being human. To take an example, the notion of 'God's people' adopts a radical shift between the Old Testament and New Testament. They were the inheritors and enactors of his promise and command, their unity symbolised by the place of the temple and the code of Torah. Above all they were bound together by a remembered and enacted *story*, a cycle of falling from grace, exile and restitution.[2] The idea of such a defined community continues in the New Testament church. However, the definition is exploded to include both Jews and gentiles, and the symbolic referents shift from temple and law to the resurrected Jesus and the Spirit. In a subtle but deeply significant move, the newly constituted and gentile Christian church found itself able to adopt (or, to take a horticultural biblical metaphor, 'become grafted into') the Old Testament story of Israel rather that reject that

[1] A serious read is the first of the multiple-volume project, N.T. Wright, *The New Testament and the People of God*. Christian Origins and the Question of God, vol. 1. London: SPCK, 1992.

[2] The Adamic exile from the Garden of Eden leads eventually to the covenant with Abraham in Genesis; exile and slavery into Egypt to the Exodus and return to Canaan; the Babylonian captivity to return and the rebuilding of the temple in Jerusalem.

rich narrative tradition as a dead end.[3] They became, almost overnight, the next chapter in a long and an old story rather than the beginning of a new one.

In the same way, if one of the great narratives of the Hebrew tradition and its sources is the relation between 'people' and 'created nature' then we might expect this also to be taken up, perhaps in transformed ways, in the new movement of the early church. We might also expect it to retain its position in the background rather than the centre of the stage, but should continue to be alert to the idea that on the largest of all canvasses the human story needs to be written together with that of all nature.

In this chapter we therefore continue the 'nature trail' on which we first embarked with the Old Testament wisdom and prophecy, and continued through the cosmic panorama, pain and reconciliation within the book of Job. The very first thing we notice is that biblical talk in the New Testament about the physical world continues to associate it with pain, with the idea that something is broken, out of joint. The subjects of nature and suffering occur in similar juxtaposition as in the Old Testament. Let us take our first New Testament example from one of the most powerful and influential minds behind the New Testament writings, the former Pharisee and subsequent apostle to the Gentiles, St Paul.

St Paul's Letter to the Romans: Creation out of Joint and the Hope of Glory

The majority of the epistolary corpus in the New Testament is attributed to St Paul, whose influence on the development of early Christianity is hard to overestimate. Intellectual, theological, political and missionary zeal reach heights in the life of Paul unequalled by even those apostles who had known Jesus personally. Before his famous 'Damascus Road' experience, Saul (as he was then known) had been trained a Pharisee. The deeply fundamental and legalistic doctrine of that movement combined in Paul with a personal energy for any cause to which he was wedded. In his early years his reaction to what he saw as a corrupt and

[3] The complete rejection of the Old Testament as constitutive within the Christian church was attempted in the early centuries BCE, notably by the mid-second-century Marcionite movement, but uniformly condemned as heretical.

subversive new movement within Judaism provoked his violent persecution of the early church.

Much early twentieth-century scholarship on Paul has assumed that, in becoming a Christian, he essentially disowned his Jewish heritage, beginning anew, preaching salvation 'by grace' rather than 'by works'. More recent work has redressed the balance, finding Paul in continuity with the hope of Old Testament Judaism, perceiving the Christian church as the way to fulfil, rather than to negate, the law and prophets. At all events, after his conversion he developed the most systematic account of early Christian theology and practice we possess. Some of it is clearly highly reflective, some written 'in the saddle' in response to urgent questions and disputes from the early Christian church, as it comes into being within the early Roman Empire, from a heady mix of Mediterranean communities of Jew and gentile, rich and poor, slave and free.

The longest and most complete of Paul's accounts of early Christian theology, and especially of its radical transformation of the Jewish story in the light of Jesus, is contained in his letter to the young church in Rome, written probably around 60 AD. After an introductory preamble, Paul begins the exposition proper with a lament on the current godlessness of humanity. As it develops, the analysis explores immorality, greed, depravity and more, but its starting point is by now striking: it is none other than the human response to the natural world (Romans 1:18):

> *The wrath of God is being revealed from heaven against all the godlessness and wickedness of those who suppress the truth by their wickedness, since what may be known about God is plain to them, because God has made it plain to them. For since the creation of the world God's invisible qualities – his eternal power and divine nature – have been clearly seen, being understood from what has been made, so that all are without excuse.*

As does Genesis within the Old Testament canon, Paul starts with creation. It is a remarkable move that achieves a lot more than calling to mind the story we have been following of chaos controlled, the emergence of order and the theological lens through which the wisdom tradition sees it. For Paul's greatest insight into the consequences of the Easter events is to realise that the story once belonging to the nation of Israel 'on behalf of' the rest of the world is now exploded—'Jesus is Lord' for both Jew and gentile alike, and all are called to follow.

Immediately he comes across a problem—only the Jews have all the law and prophets telling them what obedience to a creator God means, so how can the gentile, uneducated and unaware of the tradition of Israel and without any experience of incorporation into God's people, possibly know what is the invitation before them? Paul's answer is to identify what is commonly revealed to all humankind, what can be read by any who have eyes to see: *from what has been made*. By now we are in no danger of reading into this a superficial romantic/modern emotional reaction to the beauty of sunsets, mountain landscapes or the night sky, or at least of assuming that Paul means only this. Far more must be true: he has in mind (at the very least through his own intimate familiarity with it) all the creation tradition of story and wisdom that we have encountered already, and a good deal more besides. He is writing to those who know already that the world and its people need healing, and that creation is shot through with pain—in the human world the pain of disobedience and idolatry, and in the natural world the pain of flood and drought as much as fruitfulness and plenty.

Above all, 'what has been made' cannot for Paul refer simply to a static backdrop of timeless nature. It calls up a story with a beginning (creation itself), a middle (the painful place where he and his readers are now) and an end (a future of a reconciled and reconstituted world). It is an extremely and deceptively compact statement, and it implies that Paul admits very serious theological weight to be carried by the natural world. Equally we need to avoid reading this urge to consider 'what has been made' as an argument for theism and against atheism. That would be reading back our contemporary debates into another cultural world in which the question was not 'is there a God?' but, overwhelmingly, 'which God is there?' By pointing to nature, Paul is pointing at the creating God who is above nature yet involved with its history. In particular he is certainly lifting his readers' eyes above and out of idolatry, in a similar way to Isaiah's appeal to the distinction between human artifice and nature itself.

As is so often true with Paul, we need to read him in his own wider context, especially when so much needs to be 'unpacked' from so little. Fortunately he returns to the theme of natural creation in a much more expansive way in the central eighth chapter of the letter. This is the luminous passage at the very heart of his exposition of the 'new life', famously opening with the resonant tones of 'Therefore there is now no condemnation for those who are in Christ Jesus', and finishing with one of the best loved passages of comfort in the entire New Testament: 'For I am convinced that neither death nor life, neither angels nor demons, neither the present nor the future, nor any powers, neither height nor depth, nor anything else in all creation, will be able to separate us from the love of God that is in Christ Jesus our Lord'. But between the two—acting as a sort of bridge—is a powerfully enigmatic and far less quoted passage, yet without which Paul would be stranded, as unable to reach the concluding assurances of the chapter as he could cross a gaping chasm (Romans 8:18ff.):

> *I consider our present sufferings are not worth comparing with the glory that will be revealed in us. The creation waits in eager expectation for the children of God to be revealed. For the creation was subjected to frustration, not by its own choice, but by the will of the one who subjected it, in the hope that the creation itself will be liberated from its bondage to decay and brought into the glorious freedom of the children of God.*
>
> *We know that the whole creation has been groaning as in the pains of child-birth right up to the present time. Not only so, but we ourselves, who have the firstfruits of the Spirit, groan inwardly as we wait eagerly for our adoption as children, the redemption of our bodies.*

There are of course strong resonances of the book of Job here in the parallel placement of the decay of an entire cosmos subject to frustration, and individuals who 'groan inwardly', embodying the decay of creation within their own flesh. For Paul as much as for the writer of Job, the 'problem' is not confined to the human condition, but runs through the veins of the entire physical world. As James Dunn, one of the great commentators on Paul of the last century, puts it, 'There is an out-of-sortness, a disjointedness about the created order which makes it a suitable habitation for man at odds with his creator'.[4] Also like Job, there is a narrative momentum directed at a future healing and wholeness. However, where in Job this was implicit (even when the Lord finally speaks), in Paul it is explicit from the first. The present pain is not to be interpreted as the pain of decay, disease and dissolution, but rather the pain of childbirth, the necessary process by which new life comes into being.

But what does this 'new life' look like from a perspective that takes into its purview the entire cosmos? The description (in New Testament

[4] James Dunn, Romans 1–8. In *World Biblical Commentary*, vol. 38A. Nashville: Thomas Nelson, 1988.

Greek) is notoriously difficult to translate. The frequent rendering, as in the New International Version translation quoted above, 'glorious freedom of the children of God' is not really faithful to the original, which reads in transliteration more exactly as '. . . freedom of the glory of the children of God'. But what would that mean exactly? It is easy to see why translators, scratching their heads at this point, reshuffle the pack of Greek nouns in an attempt to make some sense from Paul's tight prose. But taking the resultant distorted translation only escapes from one puzzle to land us with another: for how can nature participate in a sort of freedom that belongs not to itself but to a people, however glorious?

N.T. Wright[5] suggests that we hold on tightly both to the Greek and to the connotations of the words Paul uses, and see where the ride takes us. For 'glory' (*doxa*) of a people or of things is very much more than a crown or status here. The word carries more sense of the honour won through a right relationship with others. This can only be the relationship with creation originally intended for humankind by the God of Paul's tradition. He will certainly be thinking of the duty of care we found in Isaiah (in chapter 3), and of that prophet's future vision for a relationship between people and nature that outgrows the pain and peril of our current experience. He must have equally in mind the command to Adam in Genesis (1:28): 'Rule over the fish of the sea and the birds of the air and over every living creature that moves upon the ground'. This rule is not the exploitative one of a tyrant, but the regulatory and facilitating oversight of the supervisor, of one who is in charge because they possess the right experience and understanding, but who knows themselves also to be under authority. Elsewhere, writing to the church in Corinth, Paul opens up another window on what he means by 'glory' in comparing the *work* of the apostles' ministry with the ancient work of Moses. Both are 'glorious', but (2 Corinthians 3:9) 'if the ministry that condemns people [that of Moses, because it brought the law that Israel could not keep] is glorious, how much more glorious is the ministry that brings righteousness!' So glory is attributable to the function of healing and restoration. It is the reflection of being the right people doing the right thing in the right place. It is the sign that the 'out-of-jointness' of the world is under the care of a community dedicated to healing it.

[5] In, for example, N.T. Wright, *The Resurrection of the Son of God*. Christian Origins and the Question of God, vol. 3. London: SPCK, 2003, p. 258.

Now it is possible to clear the interpretative fog from 'the freedom of the glory of the children of God'—it really is possible that Paul had in mind a freedom that belongs properly to creation itself, and furthermore freedom that is released only when the children themselves are in a right relationship with the world. Within his theology this only happens when they are also in a right relationship with God, the subject of the entire epistle. The Job-like pattern is unmistakable: a present pain that encompasses both a flawed physicality and a darkened understanding, marked by a broken relationship with God, points to a journey to a future in which reconciliation with the one brings also the beginnings of healing the other.

This wider theatre for Paul's theology—that of creation as a whole—has been marginalised in the stormy literature of Pauline studies of the last century. There is even one line of criticism that suggests (possibly through a faulty understanding of his distinction between 'flesh' and 'Spirit') that Paul rejected the physical world as unredeemable and tainted altogether. This is very far from faithful to his writings, and to the Judaic context in which they were formulated. We have seen how the current created order holds an important place in the very message of renewal he sees as central to the Christian mission. It is in continuity with the Old Testament law and prophets, but takes new immediacy and transformed life from the Easter events. The same combination of continuity on the one hand and radical renewal on the other characterises Paul's transformation of the theme of the future of creation itself.

St Paul's Letter to the Corinthians: The Resurrected Future of Creation

We recall that the story of the physical creation told by the Old Testament prophets contained a future as well as a past and a present—that passages such as Isaiah 11 and Hosea 2 use rich metaphors to paint a picture of a renewed creation, a different way of being physical:

> *for the Earth will be full of the knowledge of the Lord as the waters cover the sea.*

The essential point to grasp is that this future, 'eschatological' vision is no less *physical* than the current world. Indeed in many ways it is more substantial, more solid.[6] The difference, as we saw in Chapter 3,

[6] This point is famously allegorised by C.S. Lewis in *The Great Divorce*. New York: Macmillan, 1946.

is fourfold: (1) that God is more closely 'bound up' with physical nature (in a new and more immediate 'covenant with the beasts of the field and the birds of the air', says Hosea); (2) that the interrelations within nature lose their predatory and exploitative nature ('*They* will neither harm nor destroy'—Isaiah); (3) that a harmonious relationship of humankind with nature is restored; and (4) that the captivity of the physical world to decay is ended. Again this future vision is not, and never was, a dissolving of the universe and its replacement with some more 'spiritual' ethereal place called 'Heaven'—no first-century Jew would have recognised a story of this kind, and nor was Paul or anyone else in the early church about to introduce one. Something far more significant happens to this vision of a new creation in the light of Jesus.

One word that would have been applicable, for Paul, to the future hope for creation that he had inherited from his community was 'resurrection'. It picks up the sense of reversal of decay, the validation of physicality as good and eternal, and resonates with thought in later Old Testament books (such as the late prophetic book of Daniel) that begin to give more concrete clothing to the idea of a physical resurrection beyond death. But, more than anything else, it is his, and the other New Testament writers', response to the resurrection of Jesus that alerts us to the radical energy of the new movement, and to the way that they saw the continuity of Jewish hope in a renewed creation transformed in discontinuous ways by Easter. Paul's conclusion to another long and highly structured letter, the first to the church in Corinth, leaves us in no doubt about the centrality of Jesus' resurrection for everything he is attempting to build by his preaching and pastoring (1 Corinthians 15:13):

> *If there is no resurrection of the dead, then not even Christ has been raised. And if Christ has not been raised, our preaching is useless and so is your faith.*

As he approaches the climax of this letter, much as he does in Romans, he passes through a discussion of transformation within physical creation itself. But this time the climax is stated in terms not of reunion with God but of victory over death (1 Corinthians 15:55, quoting Hosea 13:14):

> *Where, O death, is your victory?*
> *Where, O death, is your sting?*

The nature metaphor Paul employs in writing to the Corinthians is not, as in Romans, the groaning of childbirth. Instead Paul calls on two analogies: the planting of and growth from a seed, and a rich expansion of the theme of 'glory' that we met in the other letter (1 Corinthians 15:37):

> *And when you sow something, you do not sow the body that is to be, but a naked husk, perhaps of wheat or some other grain. But God gives it a body in accordance with his wishes, giving each of the seeds its own body. Not all flesh is the same kind of flesh, but there is one sort for humans, another sort for animals, another for birds, another for fish. There are physical objects in the heavens, and there are physical objects on the Earth; but the proper 'glory' of the heavenly ones is one kind of thing, and that of the earthly ones is another kind of thing. The 'glory' of the Sun is one kind of thing, the 'glory' of the moon is another, and the 'glory' of the stars is different again – for one star differs from another in 'glory'. So it is with the resurrection of the dead.*

The plant is unrecognisable from the seed, it is of quite different physical form, yet is a manifestation of the same organism. Paul is correcting those who pose the question 'How are the dead raised?', presumably with all the problems in mind of decayed and worm-eaten flesh that spring (over)-literally to mind. He is saying that the question is akin to asking how a seed could ever become plant—it is another example of the wrong question—the answer is that the seed does not *become* a plant, but by its planting 'is given' the body of a plant.

Similarly, in a passage reminiscent of the catalogue of creation in Genesis 1, he redefines the question in another way: 'ask me instead about the "glory" of the resurrected body' is his suggested redefinition. Note that he is *not* saying that people will become immortalised as constellations. Both stars and fish also have 'glory'. He is recalling the variety of proper stations of different things in the created order. The implication is that in a renewed creation the bodies of those beings which have the same relationship of continuity with people as plants have with the seeds that generate them will also assume a new and proper relational position within nature. And, crucially, he can say this because he believes that the process of planting and rebirth, or of resurrection itself, is no longer something postponed to an indefinite future, but a process that has already begun by the planting of one particular human 'seed' (1 Corinthians 15:49):

> *And just as we have borne the likeness of the earthly man, so shall we bear the likeness of the man from heaven.*

This is the great discontinuity, the great surprise, for the expectant community of Jews who were looking for the fulfilment of the ancient hopes of their tradition. A tiny 'bubble' of the new created world order nucleates at the first Easter. But it does not sweep all nature before it in passivity. Instead there is an implicit invitation to participate in the coming into being of renewed creation. The gospel message contains the same beckoning finger that was extended to Job on his ash heap, extending an invitation to explore the possibilities of a new nature.

St John's Gospel: Information and Transformation

Science writers wishing to grab readers' attention for accounts of the genetic code and its embodiment in DNA have often begun in the following way: 'In the beginning was the gene, . . .'. They borrow, of course, the form of one of the most celebrated, read and discussed texts ever written, the matchless opening to the Gospel according to St John:

> *In the beginning was the Word, and the Word was with God, and the Word was God. He was with God in the beginning. Through him all things were made; without him nothing was made that has been made . . .*

There is really nothing comparable to this deep, climactic yet almost monosyllabic creation story (in Greek as in English). It is the New Testament version of the creation account that we have seen taking many different forms through the tradition of law, wisdom and prophets. It draws on their emphasis of creation as ordering and containment, but does so in a way that marks a confluence of three tributaries of thought that meet in St John's mind. They stay there on their journey along the river of his theology - two are already old by the time he is writing. They are the Hebrew thought of the Old Testament and the tradition of Greek (and especially Stoic) philosophy. One is very new: the developing Christian theology towards the end of the first century AD.

The Stoic idea of *logos* (in most translations of John's Gospel rendered as 'word', but equally inherits the idea of 'ordering principle') has its origins in Heraclitus (sixth to fifth century BC), who generalised the common idea of 'word' to a wider use of *logos* that signified commonly-held reason. In the second century BC *logos* gained further weight in the works of Stoic writers as an ordering idea or principle that possessed location not only in the minds of people but within nature itself. By the time of Philo, a contemporary of early Christianity, this ordering principle was becoming personified as a 'demiurge', or divine agent of creation that acted as God's intermediary or instrument. This notion, whose roots go back at least as far as Plato,[7] arises in the history of thought from time to time as a solution to the apparent irreconcilability of a perfect and pure God with an evil and corrupt creation. By inserting a number of intermediary agents between the ultimate divine being and the physical world, a 'buffer zone' of graded imperfection is set up. This is, as we have seen, quite antithetical to Hebrew thought and belief. It also suffers from a logical problem of continuity: if it is really to be believed that the world is 'tainted' in a way that must be removed entirely from the notion of a perfect creating being, then no chain of causality, no matter how long, does the job,[8] for the break from perfection to imperfection is qualitative, not quantitative, and must occur at some point along the chain.

Logos does appear in the Old Testament, as we saw in Chapter 3, in its ancient Greek Septuagint translation of Psalm 33. In this second tributary of John's source-traditions, 'word' orders the heavens, the stars and the waters of the deep. In this regard it resembles the earlier Stoic ideas, but departs radically from the later identification with less than perfect demi-gods. The creative word is spoken from the one God at the dawn of creation. Significantly, the ordering word in creation also plays a continuing role within that psalm of opening up hope for the future of God's people under the internal and external threats that surface and resurface in the Old Testament story. This would have been a very familiar predicament to the first- and second-century churches, and especially for the Jewish members of them, for whom John was writing his gospel. For them, *logos* brought resonances of new creation just as powerfully as for their Greek fellow worshippers.

The third 'tributary' that feeds John's climactic use of *logos* was that of the early Christian theology itself. By the end of the first century a

[7] The notion of demiurge is introduced in Plato's account of the material world in his *Timaeus*.

[8] The third-century neo-platonist philosopher Plotinus presents the most sophisticated account of how to generate imperfection from perfection in his fifth *Ennead*, but does so effectively by inserting a complete discontinuity into the chain after the highest level of 'the One'.

'Christology'—a theology of the person of Christ beyond the bare historical facts of the life of Jesus—had begun to develop in the teaching of the apostles. A central theme, an inevitable consequence of the developing Trinitarian concept of God, was the special role of Christ in the creation of the world itself, rather than just in its redemption. Early Christian theology recognised that a second person of the Trinity must participate in the being of God for all of history, not just that part subsequent to the incarnation. This had previously appeared in Paul's writing; for example in the introduction to his letter to the Colossians (2:15) he writes:

> *He [Christ] is the image of the invisible God, the firstborn over all creation. For by him all things were created: things in heaven and on Earth, visible and invisible, whether thrones or powers or rulers or authorities; all things were created by him and for him.*

What John is therefore doing in the masterstroke of his gospel's prologue is to capture the common currency of Hellenistic thought, implicitly erasing any notion of an intermediate or semi-divine agent of creation that distances God from the physical world, yet equally setting his story firmly within the community of those who had sung the psalms for centuries. In the same breath he identifies the pervasive ordering *logos* of both traditions with the person of God the Son, who is of course the subject of his subsequent chapters.

John's gospel account is full of what he himself calls 'signs', a more helpful term than 'miracle', for he wants them to point away from themselves towards their meaning in the context of Jesus' life, death and resurrection. So, for example, it is the only Gospel that preserves the account of the wedding at Cana,[9] as extraordinary a story as it is an amusing one. For reasons unstated, an understocked or overthirsty reception party runs out of wine. Jesus, appealed to through the good offices of his mother, transforms the water in six huge stone jars into the best stuff yet served. As has been noted often before, the transformation of water into wine is not of itself a miraculous process: it is achieved by a collaboration of the vine, sunlight, terrain and winemakers every year in regions enjoying the right climate. The 'wonder' of the wedding miracle is the compression of a natural process into an instant under Jesus' command, starting with inadequate ingredients.

[9] John 2:1–11

This first of John's 'signs' is surely on the same road that begins with the creative *logos*, and will end with the ultimate defeat of disorder and decay that is resurrection. His editorial comment following the Cana episode is a strong guide to the way his early Christian readers should understand Jesus' 'nature signs', 'He thus revealed his glory and his disciples put their faith in him'. If we recall the rich significance of the New Testament idea of 'glory' from our glimpse of Paul's use of the term to the Romans, then we will pick up the *relational* statement John is making. Jesus is standing in a balanced and healthy relation to the physical world: it responds to his command with fruitfulness, but his command is not out of tune with its nature.

St John's Revelation: Finally Taking Leave of Pain

We have remarked throughout our survey of biblical material that ponders our relation with 'the inhuman otherness of matter' how it does so time and again in the context of pain. The Psalms sing of it when they need assurance that the Lord of creation has not, in spite of appearances, abandoned his created people. Proverbs unearths wisdom from the formative moments of the world as a force to counter the mistaken and harmful actions the writer sees everywhere dismembering society. Prophets warn of the unravelling of the physical environment if a harmonious relationship with people is not maintained. The first two chapters of Genesis with their creation accounts preface the briars, thorns and birth pains of the third. Job brings into almost unbearable tension the soaring heights of Old Testament nature contemplation with an unmatched portrayal of physical and mental anguish. Paul weaves an explicit story of pain into his narrative of future hope unleashed by the resurrection. Pain has been our companion in the shadows of our journey throughout.

But there is one place in the Bible—I am tempted to claim just one alone—where pain is absent from talk of the physical structure of nature. And there, for the first time, it is explicitly banished. The final book of the New Testament, as the early church arranged it, is the strange Revelation to St John. Its genre is apocalyptic, which we have noted before is an ancient form unfamiliar to readers today. Throughout its central chapters the vision presented to the writer invokes lurid images of dragons, trumpets and beasts within a complex numerology. It is all too easy to misread or over-read the coded language (many commentators have done so over two millennia, and still do). This is not the place to comment on the, mostly political, referents of that apocalyptic imagery.[10] But just as the book of Job sets up parallel theatres of events—one legal and moral, the other natural and physical—so towards the close of Revelation the swirling mists of symbols clears and the writer describes a vision of a renewed created order (21:1):

> *Then I saw 'a new heaven and a new Earth', for the first heaven and the first Earth had passed away, and there was no longer any sea. I saw the Holy City, the new Jerusalem, coming down out of heaven from God, prepared as a bride beautifully dressed for her husband. And I heard a loud voice from the throne saying, 'Look! God's dwelling place is now among the people, and he will dwell with them. They will be his people, and God himself will be with them and be their God. He will wipe every tear from their eyes. There will be no more death or mourning or crying or pain, for the old order of things has passed away.'*

The passage twice quotes from Isaiah's distant vision of a reconciled world, but now the distant and veiled hope of the prophet is realised and vibrant in the vision before the seer. Alert to the symbolism of 'sea' in biblical creation stories as a symbol of threatening chaos, needing to be kept at bay beyond its boundaries, we are struck by the first radically new feature of the new 'order of things'. There is no need for the familiar boundaries of the Old Testament tradition. The 'channels' of Job's lightning and floods, as much as the threats that needed 'channelling' in the old order, are simply not present in the new.

A second remarkable component of the text is the description of the new world's advent: people are not taken up into it—the dynamic is exactly the *opposite* of a childish notion of 'going to Heaven'—the new order *descends* and seems to overwhelm the old. The sense of place, however, is preserved—the renewed Earth is in some sense commensurate with the present order. This generates different consequences for the way in which the community that embeds such future stories values and works with the Earth. If transformation, rather than destruction, awaits the natural order, then the material matters.

[10] Revelation is a book that requires the reader to work with a good commentary to make progress, but it is extremely rewarding work. See, for example, G.K. Beale, *The Book of Revelation*. New International Greek Testament Commentary. Grand Rapids, MI: Eerdmans, 1998, or, for a succinct and thoughtful introduction to the translated text, Michael Wilcock, *The Message of Revelation*. Nottingham: Inter-varsity Press, 1991.

The third fresh aspect to Revelation is the much closer, more visible co-dwelling of God with the recreated world than, by implication, is the case before. Finally, and written as if it were an inevitable consequence of the first three, is the explicit absence of pain, physical or mental. It is as forgotten, as erased from the picture, as are the forces of chaos themselves. If any reader were still tempted to retro-project an insubstantial and non-physical nature onto the vision, as generations have mistakenly done, then reading on a few verses to a detailed description of the walls and foundations of the new Jerusalem will dispel them:

> *The wall was made of jasper, and the city of pure gold, as pure as glass. The foundations of the city walls were decorated with every kind of precious stone. The first foundation was jasper, the second sapphire, the third agate, the fourth emerald, the fifth onyx, the sixth ruby, the seventh crysolite, the eighth beryl, the ninth topaz, the tenth turquoise, the eleventh jacinth, and the twelfth amethyst. . . .*

This is a scintillatingly detailed image. The choice of precious stones is an especially visual one: they all have optical properties that generate unusual refraction of light. The impression is one of overwhelming illuminated solidity (and elsewhere in the chapter we are given numerical—and vast—measurements of the fully three-dimensional structure).

But more is true of St John's vision than a reinforced physicality of the new created world. The picture also contains a strong statement about people's relation to it. Think, for a moment, of the point of view implied by the text above. Foundations are buried and out of sight, they are part of the hidden substructure of buildings that gives them strength. Yet in this final vision of the New Testament these foundations are visible at the level of their detailed structure and composition. We have met this sort of physical perspicacity before—in the hymn to wisdom of Job chapter 28, the scene opened with another set of jewelled foundations open to the eyes of human observers. Those foundations belonged to the mountains, these to a city; those minerals were silver, gold and lapis lazuli, these more varied, but the viewpoint is the same, as is the special sense of vision at work. Our eyes are treated to sight 'below the surface' of creation. The difference is that now we observe the new creation rather than the old, and can do so without constructing mines. We recall that the point of the hymn was to locate wisdom, and that it found it in 'the fear of the Lord' *because* it was the Lord who 'looked to the ends of the Earth, and beheld everything under the heavens'. This is now, just for a glimpsed moment, the perspective of John and his readers. The same order of world that has banished chaos and pain together has also managed to embody a complete understanding of its structure in the eyes of the people who behold it. Without invoking the word itself, Revelation has encoded the end of the search for wisdom, and the New Testament closes with a physical nature open to the deep view of a fully developed human perception. The dream seen at a distance from Psalms and Job becomes a present reality. The perspective is no longer one of ignorance, the world no longer threatening and the viewer no longer fearful. Far from a final closure, however, the images of a massive city, gardens and fruitful trees—a synthesis of artificial and organic activity—suggest yet more story to follow.

7

A Theology of Science?

The path we have been following is taking us to some surprising places. We have approached the task of thinking about science and theology from points of view within both projects, attempting to experience the stories of each from as close a range as possible. Working from such 'internal' perspectives, yet held closely together, casts a very different light on what is happening when we do science or theology than does a remote 'arm's length' discussion. Another advantage of such close reading is that we are less prone to oversimplification when we talk about either discipline. At one extreme, we have found no room for a clinical, monolithic scientific methodology of established fact and proof, to the exclusion of the human values of doubt, faith and belief.[1] At the other we can no longer admit a caricature of theology as the propagation of unevidenced dogma, uncritical and unchanging. Both are deeply human stories, both with a long history, and both insist on talking about each other.

So, a more significant consequence of our close perspective is a new ability to change what we might call the 'geometry' of the debate between these two communities of practice. From a safe distance it is easy to create disjoint mental maps of science and theology, each dense with connections to its own questions, communities, methods, goals and histories, but only tenuously and distantly related to the other. Think perhaps of a road map of two cities busy with streets and avenues within their own boundaries, but linked together by only a long and winding country road. In spite of the multitude of historical, cultural and anthropological reasons to explore science and theology as part of a single cultural 'city', the assumption that they belong at best only remotely connected is very rarely challenged. Rather, an extreme

[1] The bankruptcy of a 'value-free' scientific epistemology has been argued many times before. A classic exposition, still resonant with scientific practice today, is Michael Polanyi's *Personal Knowledge: Towards a Post-Critical Philosophy*. Chicago: University of Chicago Press, 1962.

disciplinary distance is taken as a starting point, both by writers who would push them further apart as well as by those who would hope to bring them closer.

There are three common arguments which take such a geometrically distant thought world as their point of departure. The first finds it natural to set up a competition between science and theology in terms of their explanatory power of the phenomena we experience in nature. It then makes an appeal—that we should sign up to one sphere of activity or the other, but never to both. So, for example, Richard Dawkins' characterisation of religion: 'Religion is about turning untested belief into unshakable truth through the power of institutions and the passage of time'[2] is deliberately contrasted with a science that tells the truth about the universe—'Religions do make claims about the universe—the same kinds of claims that scientists make, except they're usually false'. This approach is highly problematic for many reasons, not the least of which is that, historically speaking, the claims that scientists make are also 'usually false'. Good science is arguably about being false in a constructive way that takes us nearer to truth, rather than capturing truth in some timeless way. Seen in the light of Einstein's relativity, Newton was 'wrong', but we do not discard his achievement for that reason.

Science also requires rhetoric, advocacy in the face of apparent initial refutation, defending a weak conclusion or partially developed understanding in its early life, as philosopher Paul Feyerabend famously (and controversially) showed in his *Against Method*.[3] New ideas in science would die at birth without these social instruments to keep them in circulation until they gather strength of their own. 'Conflict' models of science and theology are also in danger of making as profoundly misleading representations of theology as they do of science—as we have seen from our examination of some of the biblical and historical material. There is very much more in that tradition urging human exploration of the natural world than we might expect from some representations of a biblical world-view. There is also much less written there about what we might find out from such a project; the text contains many more questions than it does answers.

The second route that takes a disjoint starting point for science and theology attempts to circumscribe their separate domains of validity,

[2] Richard Dawkins, *The Root of All Evil*. Channel 4, UK 2006.

[3] P. Feyerabend, *Against Method*. London: New Left Books, 1975.

seeking legitimacy for a geometry of thought that keeps them well apart. Most famously advocated by the naturalist Stephen Gould, his 'non-overlapping *magisteria*', or NOMA, attempted to avoid the conflict model by finding a way to sustain instead a peaceful coexistence. A 'magisterium' is for Gould a realm of validity over which a discipline carries authority (to take a trivial example, the set of all postage stamps, their design, manufacture and use would be the magisterium of philately—but it would be unwise to extend the authority of that discipline to international politics or the design of bridges). For Gould the 'magisterium' of science covers 'the empirical realm: what the Universe is made of (fact) and why does it work in this way (theory). The magisterium of religion extends over questions of ultimate meaning and moral value. These two magisteria do not overlap, nor do they encompass all inquiry.'[4]

The problem with this initially beguiling solution is the assumption that all disciplines act as magisteria (in the way that philately clearly does). The strand of truth within the 'conflict' route is the observation that both science and theology do indeed take the entirety of nature as a fit subject of their narratives. There really is only one world, and our minds are the locus of both meaning and explanation within it. So any structure or process within the natural world is a legitimate object for science, including our minds, emotions, curiosity and creativity. However, a theological tradition is also all-encompassing. In Judaism and Christianity the universe itself carries theological weight as God's creation, it carries relational weight as our human environment—with both positive and painful consequences. Keeping science and theology at arm's length artificially limits their domains of discussion—and this is inconsistent with the range of both of them.

This is why, for example, science cannot be value-free, nor a discussion of values science-free. We need to know both why we do science and how we should regulate and support it. If we attempt to disconnect science from the stories that bind our cultures together then such governance is doomed to failure, and will instead respond to purely instrumental and immediate political agendas (think of the statement on scientific research we noted in Chapter 1 from the 1993 UK White Paper on science and technology).

[4] S.J. Gould, *Rocks of Ages: Science and Religion in the Fullness of Life*. New York: Ballantine Books, 2002.

History also speaks against a model of overlapping coexistence as much as it does against the first departure from the geometry of two disjoint cities, towards conflict. Would Grosseteste, or Faraday for that matter, have recognised as valid such a separation? Faraday recorded the reason for his hope that science's search for explanations (in his case of the electricity and magnetism) would be rewarded: 'I believe that the invisible things of HIM from the creation of the world are clearly seen'.[5] The argument we met with from St Paul in the *Letter to the Romans* becomes for Faraday the grounds and justification of his research programme.

A third tempting argument criticises the approach of non-overlapping magisteria, yet equally takes its point of departure from a disjoint relation of science and theology. It attempts reconciliation by comparative methodology, while keeping the objects of enquiry distinct. A standard-bearer on this road is theoretical physicist-turned-priest John Polkinghorne, who portrays the disciplines as parallel epistemologies whose methods, time scales and subjects are different, but which can be mapped one onto the other: 'If the physicists seem to achieve their ends more successfully than the theologians, that is simply a reflection of how much easier science is than theology'.[6] As we will see in the following chapter, there is much to commend a parallel consideration of the processes involved in theology and science, but this direction of departure suffers from the same problems of dualism as the first two programmes. It allows methodologies to overlap, but not their objects of enquiry. This has the unavoidable consequence of reducing the universal scope of both narratives. Now channelled by weaknesses of its starting point, this approach begins to bring theological stories into play within the physical processes of nature, but then tends to become trapped in orbit around lengthy argumentation over God's action in the universe. So, for example, Polkinghorne pleads for 'divine interaction hidden within the cloudiness of unpredictable process'.[7] This is in danger of becoming an example of the head-on collision that Richard Dawkins claims to be the inevitable consequence of two traditions thundering towards the same territory at full speed—to coexist in this mode each looks for gaps in the other in which

[5] M. Faraday, quoted in G. Cantor, *Michael Faraday, Sandemanian and Scientist*. London: Macmillan, 1991.

[6] J. Polkinghorne, *Exploring Reality: The Intertwining of Science & Religion*. London: SPCK, 2005.

[7] J. Polkinghorne, *Science and Christian Belief*. London: SPCK, 1994.

to insinuate itself. A separation inconsistent with the scope of both persists, but now at a smaller scale: a segregation of neighbourhoods rather than of cities.

The problem with all three positions (conflict, non-overlapping magisteria and parallel methodologies) lurks in the most apparently insignificant word of any discussion around 'theology and science'. That humble conjunction—'and'—disguises whole histories of assumptions that need to be challenged. It tips us towards setting up the disjoint geometry of circles containing the worlds of theology and science, and sets us above the plane of both of them as observers and judges. The implicit assumption is that within this abstracted viewpoint we have a sufficient framework of analytical tools and a sufficiently rich thought world to resolve the tangle of apparent claims to truth, human value, of the 'clamour of voices' that cry contradiction, coexistence or resolution with equal shrillness. But does such a detached intellectual eyrie really exist? When we are working inside neither science nor theology, but examining them and their relationship from outside both, what are we really doing? Where are we? If both claim to be all-encompassing accounts of the same world then how can both be treated self-consistently in a way that bypasses the tools and patterns of thought that they themselves both claim to contain?

There is a way to avoid pretending that an adequately supported independent viewpoint exists, and to change the geometry of our thinking from the sterile, disjoint pair of systems: it is to replace the 'and' in 'theology and science' with an 'of'. A 'theology of science' generates a radical viewpoint, if a highly unfashionable one, but with the great advantage that it is self-consistent. The theological story that starts with a creating person needs to be able to speak about everything, if it is to speak about anything. In particular it can speak about the physical universe, and of human minds, and of the relation between the two. It can speak of how that special story, the one we now call 'science', belongs within the larger theological narrative of creation, pain, and healing. It can talk of what being human in an inhuman universe means, and it can do so by referring to the categories of value and purpose that constitute its natural vocabulary.

The replacement of 'and' with 'of' is perfectly consistent the other way around as well. From one point of view, Daniel Dennett's book *Breaking the Spell* is both a call for and a development of a 'science of theology' that itself has a long anthropological and philosophical tradition.[8] Evolutionary anthropology has long sought explanations for religious beliefs in terms of the social benefits that they confer.[9] Early humankind might well have benefitted from the stability of behavioural patterns and social cohesion endowed by religious tradition, which ultimately would have led to a survival advantage. From an entirely different disciplinary direction, recent neuroscience has explored the spatial activity patterns of the brain involved in religious activities such as prayer.[10] This is worthy and important scientific work—and there are no surprises that our religious traditions are supported by characteristic social and mental activity (all human activity rests on these supports). Understanding how our human awareness itself arises from the 'inhuman otherness of matter', even when—especially when—that other matter is the substrate for our own minds, is a vital task for our science.

But this is at most half the story: we need just as much a theology of science as we do a science of theology. This is true whether or not one personally chooses to explore life from a theistic, atheistic or agnostic point of view, although a theistic belief in God adds an urgent edge to the task. A project that employs an 'of' to conjoin science and theology rather than an 'and' works self-consistently with, and within, both activities. We need to know why we are doing science, not just in anthropological or neurological terms, but where science belongs in the stories we tell of our history, hopes and values, and ultimately of our purpose. Those are the theological stories.

By now the objection of anachronism—that science is simply too young to have anything to do with the millennia-old theological traditions—should no longer be an issue. We have visited sufficient evidence to show that the deep impulse to understand the physical world, to get under its skin and reconstruct it mentally, is as old as any written record of human culture. We have encountered the explicitly theological thought world of several of the most creative minds in the development of the science we possess today. Post-enlightenment scientific method lies in continuity

[8] Daniel C. Dennett, *Breaking the Spell: Religion as a Natural Phenomenon*. London: Allen Lane, 2006.

[9] The classic statement and development of this position and its methodology is due to Émile Durkheim, *The Elementary Forms of the Religious Life*. Transl. Carol Cosman. Oxford World Classics. Oxford: Oxford University Press, 1912.

[10] See, for example, Andrew Newberg et al., *Why God Won't Go Away: Brain Science and the Biology of Belief*. New York: Random House, 2002.

with this long story of 'love of wisdom of natural things', not in opposition to it. Furthermore, the experienced story of science today shares some uncanny parallels with the relationship between humankind and our physical world developed in the Old and New Testament.

Let us draw some threads from both stories alongside each other. There are a number of strongly resonant themes shared by the project of science as we experience it today, and by the theological story of our engagement with nature. It is worth examining each in turn, recapping the supporting elements of our theological and scientific journeys, before we make an attempt at teasing out what role science might be playing within a theological story.

Linear History

Both a long narrative of science and the biblical story of our relationship with nature agree that we start from a position of ignorance and, driven by seemingly unreasonable hope, work to achieve an ever-deeper understanding. To do so, both need to assume a linear history with a past, present and a future. We have thought through an interpretation of the great question-catalogue in the book of Job which implies just such a hope, in the face of unreasonable odds. As Old Testament gives way to New the momentum of story continues, with 'all creation groaning' in the hope and need of renewal. The ultimate vision of a healed world, in Revelation, is concrete and physical, with an intricate structure that is both precious and perceived. The task of eroding ignorance about the world becomes a project that humanity is compelled to adopt. Science agrees, but tends itself to be silent on just what the purpose of this project might be.[11] Even the biblical wisdom tradition is coy about the reason for engaging our minds with nature—one strand of thought (in Francis Bacon in early modern times and Gregory of Nyssa in the Patristic era) hints at the task as a reversal of the ignorance and clouded understanding introduced at the 'fall'.[12] The immensity of such a task calls for a lifetime as long as the human story itself.

[11] We should not mistake the important but very different advantages of technology for those of science, however regularly they are confused, especially by governments and funding bodies.

[12] Peter Harrison, *The Fall of Man and the Foundations of Science*. Cambridge: Cambridge University Press, 2007.

Science is replete with long-term visions. Physicists dream about a 'grand unified theory' of all fundamental forces in the universe, but there was a time when we knew about none of them. Neuroscientists talk about the distant vision of a complete map of the connections in a human brain, but before neuronal cells were identified there were no connections to speak about. Astrobiologists (and a lot of other people besides) long for evidence of life elsewhere in the cosmos, or compelling reasons for its absence, but before radio astronomy opened up its new window onto the sky there was no information on even the simplest organic molecules in space. Earth scientists pose great questions around the chemistry and structure of the crust and its continual turning over through the drift of the tectonic plates, as well as on the origins of the rocks, the atmosphere we breathe today and of life on the planet itself. Yet within living memory the idea that the Earth's crust could be in motion was simply laughable. Chemical science has long sought to move beyond the empirical to a predictive understanding of reaction, catalysis and the stability of compounds, but before the molecular theory of matter was finally established there was no avenue open to hope for such a future. Physiology marvels at the developmental pathway of organisms, whose complex physical structure self-assembles in response to the common code of DNA, whether microbe, plant or animal, but before the identification of the genetic code encapsulated in the giant DNA molecules there was no basis to understand what chemical triggers might operate in the process of forming cells. Every science has its great questions that drive a direction from ignorance and puzzlement towards dawning understanding.

The nature wisdom tradition in the Bible is directional in the same sense. It is the task of this chapter to make an attempt to frame a purpose behind these great questions consistent with the tradition.

Human Aptitude

The image in the 'hymn to wisdom' of Job 28 of the miner gazing from beneath at the 'roots of the mountains', with an eye able to perceive the world like that of no other animal, tells a particular story of human potential. Although the book of Job is the least anthropocentric of ancient cosmologies, passing by humans to exalt the great Leviathan as the pinnacle of creation, still mankind retains a special place in regard to the potential possession of wisdom. Our examination of the Lord's answer to Job has identified the same unreasonable optimism that the great wisdom questions of nature can one day be answered. For St Paul, the appreciation of the structure of creation was the starting point for a theological narrative that broke the bounds of any particular culture.

The 'seemingly unreasonable hope', which acts as an engine within the linear history of science, has been remarked on many times by reflective scientists. Albert Einstein famously observed that 'The most inexplicable thing about science is that it is explicable', capturing the continued experience of surprise that the human mind is able to interrogate the physical universe and represent it in a way that perceives anything of its hidden structure and workings. Why should creatures with minds adapted for hunting and gathering, minds that had evolved abilities to create tools, communicate socially and cooperate, be able to turn these same minds to uncovering the world below of molecules, photons and quarks, and the world above of planets, stars and galaxies?

Another theoretical physicist, Eugene Wigner, who made fundamental contributions to the quantum theory of molecular structure, wrote a seminal reflection on an aspect of this shocking ability which he termed the 'unreasonable effectiveness of mathematics' in describing the laws of physics. It is an extraordinary thing that a symbolic system that allows us to work with numbers in increasingly sophisticated ways serves as a set of mental tools with which to build working models of nature. In a beautifully reciprocal way, nature prompts new mathematics in turn; we have seen examples in the 'chaos theory' inspired by the double pendulum and other dynamic systems of higher complexity such as the Earth's atmosphere, and in the massive 'sums over histories' that lie at the heart of statistical mechanics, inspired in turn by the phenomena of boiling and freezing.

There are many instances of 'pure mathematics', developed with no idea of application to science, later becoming essential to the formulation of laws of physics. Perhaps Einstein had the case of Riemannian geometry in mind when he mused on our mental connection with the physical world. As he developed his theory of gravity that we now call 'general relativity' between 1905 and 1915, he realised that he would need to learn some mathematics worked out by Carl Riemann some 70 years before. Riemann was fascinated by the idea of curvature, and inspired to find out how our everyday notion of geometry—distances, lines, angles, areas—in three rectilinear dimensions would generalise in 'curved spaces'. One-dimensional curves can writhe and loop in a two-dimensional plane, surfaces can bulge, twist and crumple in three-dimensional spaces. How might we think of curvature in purely mathematical spaces of higher dimensions than three? A mathematical structure of great beauty emerged, containing many surprises such as the possibility of defining curvature locally, 'intrinsically' within the curved space, rather than needing somehow to view the space from 'outside'.

Einstein, two generations later, found that the physics of relativity was urging on him a necessary conflation of time, and the three dimensions of space, to think of the universe as a four-dimensional object of a single 'space–time'. Furthermore, gravity would appear in this view as a curvature of space–time generated locally by matter, like a tent roof (in two dimensions) curves away from a pole holing it up at its centre. The intrinsic measures of curvature of Riemann's mathematics became the key to making physical measurements of the curvature of our own universe. So Riemannian geometry evolved in a few short years from an abstract piece of formal, if elegant, mathematics to a set of working concepts with which to build a model of the cosmos. Wigner posed, but did not resolve, the alluring and beautiful puzzle that generalises from examples such as this. The astonishing power of mathematics sits within Einstein's still larger mystery of 'inexplicable explanation'.

The same special human aptitude for seeing into the natural world like no other creature does, and finding an extraordinary resonance when we do, has, as we have seen, been a strong theme of our theological story as well.

Deep Wisdom

Both nature-theology and science look at the this evolving understanding as decidedly more than a superficial accumulation of phenomena. We have found threading throughout the tradition high value placed on knowledge of the hidden structure of the world, some of the texts going so far as to identify such an internal grasp of nature with the notion of wisdom itself. Observation, cataloguing, classification, taxonomies: these are all essential to the collection and ordering of knowledge, but, from the earliest records we have found, constitute just a starting point for a journey from knowing what nature does to how it happens, even to the extent that predictions can be made of future events.

The double Hebraic encouragement to search for 'wisdom and understanding' (the practical and cognitive aspects of knowing) that we have met in Job and elsewhere resonates strongly with the dual methodology of experiment and theory, and with the deeply-lying longing for wisdom that their meeting can release. Other dualities mark the evolution of science—the twin seventeenth-century strands of empiricism and idealism pattern the early history of the Enlightenment. In a similar way the double tradition from Plato and Aristotle shapes the approach to nature of classical antiquity. It speaks of a deeper significance to understanding nature than simply knowing things.

Although the form of experimental science developed since the early seventeenth century is one of the special ingredients to the explosion of scientific understanding since then, it is in palpable continuity with much earlier scientific writing: the delightful manipulation of the trapped air under water recorded by Gregory of Nyssa in the fourth century, or with the material exploration of coloured light by Robert Grosseteste in the thirteenth. It is also in continuity with the notion of value accorded by people ancient and modern who have contemplated nature with a view to understanding it. 'Do you have the wisdom to count the clouds', asks the Lord of Job in the survey of great questions. We ought not to miss the implication that 'counting the clouds' would be a wise thing to do, whether we can or not. Neither should we fail to understand that it is also wise whether it would be 'useful' or not. Scientists know that the value of what they do wells up from far deeper sources than the instrumental or economic, but they find it challenging to explain where this overwhelming intrinsic worth of science comes from.

Ernest Rutherford was a gruff New Zealand-born physicist and Nobel Laureate who in the early twentieth century pioneered nuclear physics in the UK in Manchester and Cambridge. When he quipped that 'All science is either physics or stamp collecting', he was being very unfair to biology and chemistry, and certainly overclaiming for physics, but was articulating the drive of science that goes deeper than an account of the phenomena themselves. He devised a famously simple but brilliant experiment that detected the paths of fast-moving charged 'alpha particles'[13] as they passed through a thin foil of gold. The astonishing observation that a small fraction of them bounced back from the flimsy sheet

[13] Equivalently the nuclei of helium atoms, consisting of two protons and two neutrons.

became hard evidence of a radically new conception of the structure of the atom itself. Accompanying a deep physical insight with a beautiful piece of mathematics, Rutherford showed that the splay of directions in which the particles were scattered by gold could be explained if the vast majority of an atom's mass was concentrated into a tiny 'nucleus' at its centre. The experiment and its reflective analysis completely transformed our perception of the structure of atoms. Such imaginative leaps of understanding leads to new ways of deep 'seeing' that take us into the innermost structures of the natural world.

The Rutherford experiment is a transparent illustration of the dual methodology of experiment and theory within the science community. It is perhaps a pity that the drive towards specialisation has led more often than not to scientists describing themselves today as either 'theoreticians' or 'experimentalists'. The reward of specialisation is of course the greatest possible expertise in techniques of mathematical reasoning or computer calculation in one case, or the use of advanced microscopy, laser physics or sophisticated chemical reactions in the other. But the most effective researchers today keep very firmly aware of both the practical wisdom of intricate manipulation of matter and the contemplative understanding of theory. The new ways of 'seeing', which both guide experiment and emerge from it, require the dual harnessing of theory and experiment.

When they do unite in a spark of insight, and a new light shines on a previously dark part of the unknown world, something almost unspeakable happens. Scientists only very rarely write this down—it is almost too personal. But it is what we work for. Those miraculous moments when the fog clears and we know something for the first time really are 'more precious than rubies' to borrow biblical language. A conversation I was recently enjoying with a very serious theoretical physicist wandered into this territory. Perhaps because the rules of the conference we were participating in disallowed any attribution of statements made there, he felt freer than usual to talk about the intensely personal and frankly ecstatic moments of achievement (or are they moments of gift?) at which a physical principle at work is understood for the first time. Richard Feynman is one of the few who have written about the experience of realising a previously-unknown law of physics and even being aware that one is the first person in history to have understood it. The deeply cherished and valued nature of these unsurpassed experiences is a sign of how deep is the human longing that they meet.

A senior American politician once asked the director of a national particle physics facility what contribution his laboratory made to the defence of the USA. 'None at all, sir', was the refreshingly honest reply, 'it is simply one of those things that makes the United States worth defending'.

The Ambiguity of Problems and Pain

We have remarked before that at each point where biblical thought turns to contemplation of the natural world there is an accompanying wince of pain; until, that is, at the very end. Both of the traditions we have explored live within the ambiguous relationship between humankind and the natural world. Just as we might be surprised, on reflection, that our minds are able to re-conceive the laws of the physical world, we might be equally shocked that we find it necessary to do so. By this I do not refer to the necessity of technology, it is rather the experience that we find ignorance of the world uncomfortable or frightening. We are behaving like fish out of water when we start to worry about the 'inhuman otherness of matter' with George Steiner, for when we look around find ourselves embedded within this very matter, indeed constructed out of it from our sinews and bone to the delicate brain tissue that supports our thinking. Why then are we discomforted in a material world?

Our theological stories weave into themselves a strong thread of the relational ambiguity between humans and nature. Genesis 3 draws on the metaphors of thorns and the pain of childbirth to create an expectation that nature and natural processes will continue as a source of human pain. Isaiah counters with the picture of the knowledgeable farmer, in an experienced relationship with the land, exercising wisdom in its husbandry. Jeremiah complicates the prophetic response to Genesis with his lurid picture of creation rolled back into chaos once more if this wisdom is rejected. Job, his own body shot through with pain and disease, embarks on a quest for justice and a long debate with the four friends, eventually bringing the questions of pain and of the inhuman otherness of matter into the same focus. There is no current resolution to this 'groaning of all creation' as Paul calls it when he needs to throw the arms of his argument for resurrection hope around the whole world. The response urged on his hearers, living in the 'now', is to understand better why there are unresolved clashes of expectation and reality, of desire and darkness, and to begin an incremental journey towards healing and understanding, rather than to expect a sudden resolution of suffering. We might be reminded of Job's silence with regard to pleas that he be healed and restored; more important to him is the need to understand and to be vindicated in his predicament. So a common motif to both Old and New Testaments is the call to wait,[14] of the resistance to impatience, of a kind of reconciliation with time.

George Steiner, in his search for the place of art and literature in a recovery of meaning,[15] draws, surprisingly, on a Christian metaphor to express the common encounter of the extended time between loss and reconciliation; he calls this time 'Saturday'. This is of course the central, inner movement of the Easter story from Good Friday to Easter Saturday, but Steiner appeals to the universality of the experience. He refocuses on our current predicament as lying, not with the more singular events of crucifixion and resurrection, but with the day in between, the day of patient waiting. All broken relationships, all loss, bring not just current pain. A good part of the heaviness comes from the knowledge that there is pain still to come, that the time of waiting for healing is also a time of groaning.

At this point it is worth taking a reprise from Chapter 1 of the endearing contemporary account of a sudden realisation of the gulf between our minds and the world we observe, given by the author Bill Bryson in the preface of his excellent historical account of science *A Short History of Nearly Everything*.[16] Bryson, an arts graduate, journalist and former Chancellor of Durham University, for whom science was peripheral for most of his working life, gazes out from an aircraft window and begins to wonder at the formations and patterns of the clouds he sees far below, and the rich and deeper blue of the sky above. Seized by a creeping horror that he knows nothing about any of these things, or indeed about a thousand other questions, he conceives the idea of finding out—whence the genesis of his book. It is an outstanding account of a latent and suddenly revealed desire to understand something of the natural world in response to a felt conceptual alienation from it (and incidentally a delicious read—Bryson was even more interested in how we know things and who was responsible for finding it all out, than just what we know).

[14] Admirably handled in W.H. Vanstone, *The Stature of Waiting*. London: Longman and Todd, 1982.

[15] George Steiner, *Real Presences*. London: Faber, 1989.

[16] Bill Bryson, *A Short History of Nearly Everything*. London: Transworld, 2003.

Painful ignorance is felt within the process and community of science too. We heard a contemporary story of a group of scientists faced with the frustration of an experiment 'that went wrong' in one of our stories of science in Chapter 2. Sometimes the pain fails to resolve, even in the face of understanding. The story of chaotic dynamics illustrates how we can learn to live with ignorance, even to make use of it in statistical theories of complex matter. But our science is not simply the echo of our own minds, and sometimes grates painfully against our expectations. Einstein famously railed against the conclusions of quantum mechanics, the counter-intuitive and strange theory of elemental matter that emerged with the work of Heisenberg, Schrödinger and Dirac in the 1920s. It is worth taking a deeper look at this example of nature surprising and discomforting our expectations.

The fundamentally new notion of quantum mechanics is that a particle can be in a super-position of states at any one time, and that this super-position is physical not statistical. We are used to common ('macroscopic') objects existing at well-defined locations—we even joke about the impossible but desirable ability to 'be in two places at once'. But the atomic world really is like that. A particle is, in some measurable sense, 'in several places at once'. It is a property of the particle in the way that it evolves with time and interacts with the environment, not of our knowledge about the particle. Electrons may have two paths open to them between a source and a detector, and respond with a measurable position on one of the paths whenever they are observed on the way. Yet when not observed like this in transit, they show a distribution of arrival points that can only arise from the interference from the two distinct paths, as if each electron carried information on both paths throughout its journey, rather than travelled along just one.

Furthermore, identical experiments on individual electrons have an uncertain outcome however carefully they are carried out. The dissolution of classical physical ideas of locality, repeatability and (at least) potential knowledge of an observed system affronted Einstein, who embarked on a prolonged intellectual battle with Danish physicist Niels Bohr, trying to show that the new theory led to absurdities. 'God does not play dice!' became an Einsteinian mantra, but would be met by Bohr's 'Albert, stop telling God what to do'.[17] The tension and strain in

[17] Abraham Pais, *Subtle is the Lord: The Science and the Life of Albert Einstein*. Oxford: Oxford University Press, 1982.

the relationship of these two great scientists is a reflection of, and emerged directly from, the difficult relationship, and its growth pains, that humankind has with nature.

Perhaps the most ambivalent aspect of science, not so much painful as drawing on an extreme combination of effort and patience, is the enormity and difficulty of its task. This is the counterweight to Einstein's surprise—explicable the world may be, but only gradually does that explanation dawn after centuries of combined effort, a myriad of wrong turns and misconceptions, the devotion of years to new techniques and new mathematics that give handholds to climb higher and eyes to see further. The divisive current argument in the physics community around the mathematical structure called 'string theory' stems from the difficulty and reach of the scientific task. The theory stems from a great dream that physics believes it can allow itself. This is that, one day, we will possess a single theory of the fundamental laws of nature, consistent with every phenomenon we know about and with every experiment we can perform, mathematically expressed and internally consistent. Such an ideal is called a 'grand unified theory' because it would finish the work of bringing initially disconnected forces into the same, interconnected, framework. James Clerk Maxwell showed in the late nineteenth century how magnetism and electricity were two aspects of a single force–'electromagnetism'. The electromagnetic force was itself a century later united with the nuclear forces unknown to Maxwell. The remaining outlier is gravity—general relativity, almost as if it had inherited the recalcitrant revulsion of quantum mechanics possessed by its inventor, to this day resolutely refuses to be reconciled mathematically with the quantum mechanical nature of the other forces.

But a patchwork physics is painful—our intuition cries out that the world is cut of one cloth. Recall that this motivation was clearly at work even in the thirteenth-century early cosmological physics of Robert Grosseteste. We *ought* not to be using mathematically inconsistent theories for atoms and for galaxies, messily stitching them together along a seam where we do not look very often. One possible avenue that might lead to the hoped for reconciliation of gravity with the other forces is the web of high-dimensional quantum structures that is string theory. The name is not undeserved—the fundamental entities in string theory possess a mathematical structure that really does correspond to a tiny string rather than to a point.

Mathematically alluring though string theory is, its current inability to connect with experiment, and furthermore its perceived potential inability ever to do so, debars it for some from admission into science as such. The debate has been public and sometimes acrimonious.[18] Accusations of small-mindedness and irrelevance fly across the world's academic physics community like the old anathemas of the early church councils. There has been less reflection on the sources of discord: the long and patient climb, with its many falls, that understanding nature has always demanded, and the sheer difficulty and demand of the task, the seemingly everlasting patience required to hold together the wisps of our current knowledge. The breaking of the 'dual wisdom' of theory and experiment, and its resonance with the biblical 'wisdom and understanding', points to a deeper significance of the felt pain within this argument than a contemporary spat about methodologies.

There is yet more to the pain of our relationship with nature than the discomfort of ignorance. The world's dynamism is often destructive—the earthquake and the storm still threaten, and repeatedly overwhelm our technologies of self-preservation. The chaotic patterns of the Earth's weather system bring flood and drought as well as irrigating rain and sunshine. The biological seam of the world also menaces with infection and morbidity. Crucially, in the last millennium, humankind has moved from a minor status within the biosphere, in terms of our effect on the energy and chemical balance of atmosphere and ocean, to a dominant position. A human population growing towards 10 billion, in possession of surging energy-rich technologies, now holds the balance of climate, fresh water and crop cycle. This makes ignorance of the world even more frightening—for now we see that making the wrong move may alter radically the degree to which our planet is hospitable to human life.

Somewhere deeply within the heart of science, almost silent, are the groans of labour pains, the work to turn our present ignorance and our search for meaning into understanding. From the environment around us, whose development we are increasingly affecting, come urgent pleas to establish the 'covenant with the stones' of the wisdom literature.

[18] See, for example, Lee Smolin, *The Trouble with Physics: The Rise of String Theory, The Fall of a Science, and What Comes Next*. New York: Mariner Books, 2007.

Order and Chaos

The big themes of human suffering and the darkness of ignorance arise in part from another creative tension that has remained with us, whether we are engaging in science or in nature theology. This is the delicate and intriguing balance between order and chaos, between the regular and predictable, on the one hand, and the stochastic and the random, on the other. There are surprises here as well: the development of non-linear dynamics and its mathematical notions of chaos and unpredictability would have disappointed an eighteenth-century mechanist like Laplace, but it has led to rich new directions for physics in the twentieth and twenty-first centuries.

This paradoxical programme of science to comprehend chaos, signalled by its realisation that chaos is part of the creation of emergent structures in the world, rather than their destroyer, shares an unmissable resonance with the part played by order and chaos in the biblical narrative we have also been following. Recall the earliest strands of the creation stories in the Proverbs and Psalms: our surprising observation was that they identified the central creative act not so much the summoning of matter from nowhere and nothing but as the ordering of the elemental deeps, the chaotic threat of the waters and the drawing of boundaries that endow the cosmos with structure. Just as the fundamental creative act is one of ordering, so the fundamental energy that drives it is the 'word'. God, or Wisdom, speaks—and the floods retire, the Moon marks the seasons, the Sun rises and sets. The formal and developed account of Genesis begins with an idea of formless chaos receiving day by day the imprint of structure: we are reminded of the visual poetry of the Lord's answer to Job as a jagged horizon emerges from darkness in the morning light as if stamped out by a seal.

But the theological narrative of chaos and order is far more complex and interesting than a simple triumph of one over the other. God is not only the shaping force of order, he also unleashes (e.g. in Jeremiah) the forces of thunder, clouds, lightning and wind. There is an uncomfortable undercurrent of pain and puzzlement here. The ordered and productive work of the farmer described by Isaiah can be undone in a moment by the uncontrolled flood that breaks over the river banks, or by the ravages of forest fire. Yet these have no other source than the creator himself. The most detailed articulation of the paradox is of course woven into Job's dialogues. Job longs to understand the apparent lack of justice in his own story, but perceives it written large on the backdrop of his global theatre, where the forces of creation itself wreak destruction apparently at random. His universal perception becomes his universal projection—his planetarium roof diffuses the images of his argument. His accusation that God is out of control of creation, just as he is out of control of justice in the lives of individuals, drives the long journey through the dead-end arguments with his friends, illuminated though they are with occasional glimpses of something deeper. The forces of the wind and clouds, the lightning and flood, become recurring metaphors for the inner energy and life of the physical world that must be channelled, or given a 'way', rather than entombed or left to unbridled destruction. There is a subtle handling of 'knowledge' throughout the journey, and its distinction from its dual—'understanding'.

An example in one of the questions put to Job in 'the Lord's answer' illustrates how this idea works—*Who has the wisdom to count the clouds?* (Job 38:37). Its setting is the heart of a series of questions about floods, lightning (*'do the lightning bolts report to you "here we are"?'*), rain and the effect of all this cataclysmic meteorology on the parched ground. The section is arranged around a centrepiece question: *'who endowed the heart with wisdom or gave understanding to the mind?'* The detailed, quantitative knowledge of the paths of the lightning and the number of clouds are well-chosen examples of data on natural phenomena that by themselves do not result in understanding.

Complete knowledge is not the same thing as, and may even be antithetical to a grasp of, significance—in particular of emergent structure. We can understand the forces and balances of a climate, but this understanding does not come by counting clouds. It calls on a wider perception of phenomena, a deeper probing of them, driven by a *wisdom* that sees the relationship between phenomena separated in space and time. A wise science of nature learns to connect apparently distinct and different things. Even the juxtaposition of the topics within the Lord's answer suggests an early suspicion of hidden structure in the background of the world. Framing the 'counting the clouds' passage is the striking and enigmatic question about the 'laws of the heavens' that we dwelt on during our journey through the book—*'Do you know the laws of the heavens? Can you set up God's dominion over the earth?'* And close on its heels we find lionesses seeking prey to satisfy the hunger of their cubs. While we cannot claim any form of theory presented here of how nature works in ways that connect the laws of the heavens to the climate of the Earth, and thence to the well-being of living things, the great nature poetry of Job, as well as the traditions that lead up to it, is driven by a conviction that such connections do exist. The conviction carries two other aspects with it: on the one hand that an understanding of them is not forever out of the grasp of human minds, and on the other that the task of acquiring it can only be a long and difficult one. We call to mind not only the images of thorns and briars of the Genesis account of the human task to care for the natural world, but the intellectual balance of rationalising order and chaos.

The theoretical foundation of thermodynamics—the powerful science of matter, energy, heat and temperature—was formulated in the nineteenth century by Gibbs, Boltzmann and Maxwell. Their 'statistical mechanics' derives the properties of ensembles of very large numbers of particles by making a virtue out of the ignorance we must always have of the detailed trajectories of every particle within them. We encountered in Chapter 2 a fundamental and beautiful idea deep in the structure of statistical mechanics—the 'sum over states'. Rather than attempt to follow or predict what actually happens in a gas of trillions upon trillions of careering and colliding molecules, in a liquid of billions of molten molecular polymer strings or in a magnetic metal of as many atoms as there are stars in a million galaxies, the theory constructs instead an enormous sum over all *possible* histories of all these objects, weighting each one with a likelihood of its occurrence. The brilliant achievement of these pioneers is the general result that allows these likelihoods to be calculated. The most delightful surprise is its simplicity: each possible state of a system is visited with a likelihood that depends on the ratio of its energy and the temperature—and on nothing else.

All that we require in order for the predictions of statistical mechanics to apply to the real world is that the unpredictable dynamics of the atoms and molecules of a material explore these states in some 'representative' way. It is in fact very important that the dynamics are not regular—if they were then the sampling of configurations becomes too restricted. To understand this think again of the realm of predictability that is the system of Sun and planets: a 'statistical mechanics' of the solar system would fail because the orbits of the planets are too regular—they do not explore all the possible motions that are possible in principle (thankfully for the dwellers of planet Earth). A statistical theory would need to include in its sum the possibility that the Earth is occasionally flung into the outermost darkness beyond Neptune, tapping into a fraction of the orbital energy of Jupiter to do so.[19] We need a much more intense mixing up of motion to underpin the huge sums in statistical mechanics, and the underlying paths need to be irregular in the sense that they do not revisit identical states (as do orbits) but wander in a representative fashion throughout the space of possibilities.

The term for this rather special property is *ergodicity*. In the very few simple cases of systems in which it has been possible to prove the dynamics to be ergodic, the property emerges from the very chaotic nature that makes the trajectories themselves unpredictable—statistical mechanics and all its phenomena become possible just at the level of complexity beyond which deterministic mechanics becomes impossible. It is a matter of faith in our current intuition that in all realistic cases of complexity, far beyond the hope of a proof of ergodicity (all real applications of statistical mechanics fall into this class), that it nonetheless holds.

The extraordinary property that a pan of water, perfectly stable as a liquid at 99.9999°C, will transform into gaseous steam at 100°C would be incomprehensible without this deep perception of how the chaotic motion of vast numbers of molecules in a material conspires to explore the even vaster space of their possible configurations. Of course the phenomenon of boiling is commonplace—we are used to it, but that does nothing to explain how it arises, and why so suddenly at a sharply defined temperature. We do not 'see' why this sudden change of state should happen when we focus on single molecules. Gibbs and Boltzmann were bold enough to show that it is possible to think about immense numbers of molecules at the same time. As temperature increases by incremental amounts, nothing discontinuous happens to the typical velocities of motion of water molecules: they increase just as incrementally in proportion. The discontinuous and spontaneous switch in the emergent behaviour of the ensemble at a precise temperature is a property of the combined likelihoods of two sets of trajectories involving billions of molecules—the first describing the dense roiling around each other in the liquid state, the second the rarefied and ballistic exploration of space we call a gas. It is the billions of billions of molecules that make the transition so sharp.

[19] Actually there is no hard result in orbital mechanics that can provide us with permanent comfort against this possibility, as the planets are weakly coupled with each other's motion by gravity, but the timescale before such an event becomes at all conceivable is very long indeed.

At each temperature at which water is in liquid form there is, in the shadows, another possible set of trajectories of all its molecules that correspond to the much less dense state we call a gas. To reach this other shadowy state requires each molecule to acquire a little energy from the environment, a little 'kick', to escape from the weak bonds that hold it against its neighbours in the liquid. Tiny changes in the probability of this happening for any one molecule become multiplied over and over again through the vast numbers in even the smallest droplet—and as soon as the set of gaseous trajectories becomes at all more probable than the liquid ones, it becomes *immensely* more probable. Philip Anderson, the twentieth-century American physicist, was fond of saying of statistical mechanics, 'More is Different'.[20] Entirely new phenomena appear out of ensembles of particles, which vanish when we peer too closely at any one, just as the subject of an impressionist or pointillist painting disappears to our perception completely if we look at the canvas too closely. Standing back from swirling molecular chaos allows us to perceive, even to understand, why greater structures emerge providing, that is, we exercise the unproven but necessary faith in the science of the ergodic hypothesis.

So the underlying chaos of the molecular world turns out to be an essential, if unseen, supporting structure for all sorts of ordered phenomena. The very existence of liquid water is a commonplace example, but it is a relevant one to our existence, since, for all forms of life that we know about (at least), water constitutes an essential ingredient (and in the vast majority of cases the essential environment). That is the very first step in the link between the molecular chaos of our world and the emergent structures of complexity, and of life itself. We encountered the idea of 'self-assembly' in the case of the rod-like peptide molecules, and their formation of fibrillar structures, in Chapter 2. The theory of those structures assumes just the same methods to achieve an understanding of the spontaneous formation of the twisted microscopic bundles and fibres that we saw with the aid of an electron microscope. The mathematics simply encodes a way of calculating the likelihoods of all the constituent molecules moving independently, or together, in various forms of ensembles. Just as the gas state becomes overwhelmingly more likely

[20] Anderson published a short but very influential paper in the journal *Science* in 1972 with this title. He introduces a general way of understanding the transitions we have discussed in terms of 'broken symmetry', an idea that had previously led him to be the first to outline the mathematical structure of an example now called the 'Higgs mechanism' in particle physics.

than the liquid state of water as its boiling point, so one form of fibrillar structure dominates over all others as the concentration of the tiny peptide rods is changed.

One more example of 'impressionist' chaos underlying emergent structure is worth a mention because of its relevance to living organisms as well as its ubiquity and beauty. For its first observation we return to the eagle-eyed Robert Brown. If his name had not been attached to the random motion of small particles due to heat that allowed Einstein finally to unlock the window on the molecular world, he would surely have been better known for his observation that plants seemed to be constituted throughout in microscopic compartments he called 'cells'. The cells had well-defined walls, and shapes that corresponded to the parts of the plant they constituted. Microscopic study of tissues from both plants and animals rapidly concluded that the cellular structure of life appeared everywhere, and even provided a fundamental categorisation of the living world into the two domains of 'single-celled' and 'multi-celled' organisms. Cell walls are, perhaps surprisingly, not constructed by any systematic placing and adhering of their units, but constitute another example of 'self-assembled' structures.

Their units are molecules called 'lipids', of the same order of size as the nanoscopic peptides of the self-assembled fibres, but differing chiefly in their bipolar nature. Although there are many types, they all share a head-and-tail structure, the head predominantly attracted to environments rich in water and the tail composed of short hydrocarbon chains locally similar to the molecular make-up of oils. In order that the tail-end of the lipid is not geometrically much narrower than the head, there are in living systems always two tails attached to each head. As a consequence of the oily chemistry of the tails, these have a tendency to be repelled from water-rich environments, for the same reason that oils refuse to mix with water.

Now suppose that large numbers of such lipid molecules are produced in a watery environment (like a living organism). Initially the oily tails find themselves in a very 'high-energy' state, with unfavourable contact with water painfully common. But as the random motion ('Brownian' again) operates, they will, if concentrated enough, naturally have occasion to collide. When two lipids come into contact head to head and tail to tail, they locally mimic the favourable separation of oil and water and will thereafter tend to stay in that configuration. Other geometries of collision, such as head to tail, will generate local repulsion rather than attraction, and will not be so long-lived.

We can see what the result will be after many such collisions between lipids, and attracted conglomerates—eventually a stable assembly can be formed in which all the oily tails are shielded from the watery environment. Its structure is that of a *double layer* of lipids, each component with heads and tails aligned as in a tightly packed crowd of people. The two 'leaflets' of the double layer now arrange themselves so that the tails all face each other, away from the heads, which themselves all face the water-rich regions outside the layer.

Figure 7.1 illustrates how such a self-assembled membrane emerges naturally from the double chemical character of the lipids, and adds one more final trick that the process of self-assembly plays in its goal of removing contacts between the oily tails and water. Even the contacts at the edge of bilayer sheets can be excluded if the membrane wraps up into a shell. Then there are no edges at all, and all the tails are shielded from the water, which is now compartmentalised into internal and external regions.

This is the underlying structure of all living cell walls, though in organisms the complexity is of vaster compass, the membranes comprising typically hundreds of different types of lipids. They also contain

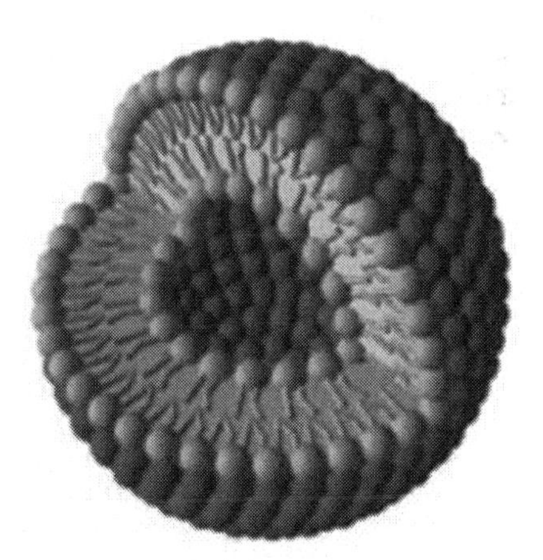

Figure 7.1 A cut-away diagram of a self-assembled vesicle whose walls are a double later of two-tailed lipids. The water-loving molecular heads (shown round) face either outwards or inwards against watery domains. The twin, oily, tails all face into the interior of the double layer forming the wall of the vesicle. Image credit: Dennis Barten/Wikimedia Commons.

more complex protein molecules that endow the whole cell wall with functions such as the controlled passage of nutrients and signalling molecules into and out of the cell. However, all this complexity is 'controlled chaos'—removing the random molecular motion is equivalent to reducing the temperature to absolute zero. No life operates there.

The lipid walls themselves become important theatres for random diffusion essential to the functions of the cell. As the Austrian physicist turned biologist Max Delbrück pointed out, if the cell requires a reaction between two molecules, then relying on Brownian diffusive motion to bring about their collision comes at the price of a long delay: random motion cannot be relied upon to select the shortest path between two starting points—far from it. But there is a deep consequence of the randomness of diffusion: the average length of the waiting time for a collision depends very sensitively on the *dimension* of the space in which the random search of one molecule for the other is embedded. The waiting time for a random diffusive search between two small objects in a two-dimensional space (such as a cell membrane) is overwhelmingly faster than in a three-dimensional space (such as the interior of a cell). The 'controlled' or 'channelled' chaos of random thermal motion on a membrane permits the vital molecular dance of life to take the floor.

We seem to hear resonances—we can hardly help it—of the Lord's answer to Job. We might rail against the chaotic and complex elements of nature that threaten our well-being or escape our understanding, but, when we respond to the invitation to peer into the beautiful structures of the natural world we are so interested in protecting, we see them built upon a microscopic world of disorder which is the substrate of life itself.

Questions

The fundamental role of questions in science and theology emerges naturally in both fields, but has not attracted the attention it might have done in the discussion between them. Surprisingly, our textual high-water mark of the extraordinary nature questions of Job 39–42 has not been drawn on significantly in the debate, in spite of the theological, as well as aesthetic, weight that it carries. This is a pity, for one of the caricatures of 'faith' painted by its more outspoken critics is that it constitutes an incurious and unquestioning acceptance of dogma. There are extreme examples of rigid faith traditions where this accusation might hold, but it entirely fails to represent faithfully the long tradition of questioning within the Bible itself, commentaries upon it since then, and the faith communities which keep those questions alive today and forge new ones. And in the case of Job, it is in the end questions, not principally the answers, that constitute the really decisive step in the healing of his own mental anguish. Not only that but, as we saw in our close reading of the whole book, the nature questions serve also as pointers forwards to a future where answers may be won, and backwards to the great question of chapter 28, 'But where can wisdom be found?'

The healthy tradition of questioning includes openness to other cultures of thought. We have already encountered a little of the alert and questioning early science in the thirteenth century. The 'high Middle Ages' are being increasingly recognised as a time of intellectual development, rather than stagnation, in Europe—in a direct line of development that leads to the renaissance and 'scientific revolution' of the seventeenth century. Arguably, the role of fresh questions set off by new collisions of ideas was central to this earlier scientific revolution. Far from the idea of a closed, defensive, dogmatically rigid world, we see thinkers like Grosseteste unproblematically bringing Christian minds to Greek pagan texts, transmitted via Muslim scholars and commentators.

To take just one example, Grosseteste was familiar (among the earliest in northern Europe) with the writings of the Spanish and North African Muslim scholar Ibn Rushd (1126–1198), known also as Averroes. The Muslim scholar's three commentaries on Aristotle's *Physics* were clearly influential in the imaginative thinking on light a century later that we examined in Chapter 2. Likewise, Averroes' early questions that approach the forces and motion of matter began to sketch the concept of inertia, later developed by Buridan and Oresme in the fourteenth century, and given solid experimental form by Galileo in the early seventeenth. Of course questions on their own are sterile until they lead to debate and disputation. Those activities are in turn only possible in an environment that is confidant in supporting opposite opinion and living with uncertainty. The Hispanic Islamic world of Averroes and the northern European culture of Grosseteste and Aquinas were clearly neither uncomfortable nor defensive in an environment of questions and disputations.

A delightful example of the heritage of the literary question form in approaching nature is found right at the start of the European 'first scientific revolution'. Adelard of Bath was a twelfth-century English scholar, widely travelled (including Sicily and Turkey) and fluent in Arabic. He was already influenced by the Islamic tradition of commentary on Greek science, although he wrote before the new influx of Aristotelian physics took hold of the European intellectual agenda. His celebrated *Quaestiones Naturales* (Questions of Nature) takes the form of a dialogue between himself and his curious yet sceptical nephew.[21] Seventy-six questions in all are posed, ranging over much the same subject landscape as Job's do. In one well-known episode, his nephew asks how the Earth can be supported, when of all objects requiring support it is the heaviest. Adelard's answer is a model of imaginative logic—explaining that 'down' for those on the scale of the surface of the Earth is really 'centripetal' at the scale of the entire planet. He helps his nephew to take a distant standpoint from the spherical Earth, and to see in his imagination that, for all observers on its surface, locally downward motions are really all convergent on the centre of the Earth. Adelard invokes the solidity and resistance of matter, its own elastic property, coupled with the centripetal tendency, to explain how the material of the earth can be self-supporting. The *Quaestiones* became a favourite text of the next two centuries. It carried with it not only a thought-provoking account of the physical world-view in 1150 but also an unstated value of questions, debate and free enquiry. By implication, Adelard stands in the tradition of Bede and subsequent Christian scientist theologians, for whom logic, not authority, was the route to understanding the workings of nature, and for whom their exploration of nature was a Christian calling, central to their vocation, neither sideline nor aberration. Questions still propel the creative energy of science.

Werner Heisenberg was one of the key thinkers at work during the forging of quantum mechanics in the early 1920s. In the space of a few intense years our fundamental conception of how matter works at the smallest of length scales was completely reshaped. The revolution was driven by a few extraordinary minds in Europe: Schrödinger in Zürich, Heisenberg and Born in Göttingen, De Broglie in Paris, Dirac in Cambridge, and their unofficial mentor-in-chief Niels Bohr in Copenhagen. So many aspects of our everyday interactions with matter were overthrown in the interval of a few years that it has been hard to know what single label to attach to the new thinking, with the result that it got stuck with a rather poor one.

[21] Adelard of Bath. Charles Burnett (ed.), *Adelard of Bath, Conversations with his Nephew: On the Same and the Different, Questions on Natural Science, and On Birds*. Cambridge: Cambridge University Press, 1998.

It is true that, in some cases, quantities that we are used to thinking of as available in any amount chosen from a continuous selection (such as energy) are in truth only available in discrete amounts, or 'quanta'. But it is rather more astonishing that some deep *logical* assumptions that had reigned for centuries were also shown to be contestable, rather than axiomatic, when applied to matter. How often in science do we discover that 'obviously true' things are quite wrong. For example, classical physics makes (quietly) the assumption that the position of an object is exclusive: the atom is *either* at position *A* or at position *B*. Quantum mechanics, as we noted above under our discussion of pain and puzzlement, requires other possibilities—for example that the state of the atom might be a 'super-position state' composed of elements of both states *A* and *B*. In an extreme limit, this 'superposition mechanics'[22] allows matter to behave like waves, focusing, reflecting and diffusing as it travels. Quantum theory also places the act of observation at its heart, rather than an optional action irrelevant to the system observed. Still puzzling in its extraordinary ramifications, the future behaviour of matter depends unavoidably on whether or not observations have been made on it at moments in the past.

An intriguing element of the story of quantum mechanics is the role of experiment. More usually in science, progress in theory and experiment go continuously hand in hand, but in this case there was (prior to the short period of major breakthroughs) no shortage of unexplained data. Experimental findings were replete with paradoxes. In particular, data on the type of light emitted from atoms (and vitally the light *not* emitted) had been carefully gathered for years, and none was explicable by any form of classical physics.[23] Commonplace notions, such as the chemical bond and its stability, were mysteries. The very solidity of matter remained unexplained (even eight centuries after Grosseteste commented on the problem of classical atoms in this context, as we saw in Chapter 2). A huge headwater of experimental observation needed to find its channel, but no amount of steady erosion of the old thinking was ever going to provide it.

[22] John Polkinghorne has suggested that quantum theory might more appropriately be called 'superposition theory', see, for example, his *Science and Theology*, London: SPCK, 1998.

[23] From the very early nineteenth century, the experimental science of 'spectroscopy' enabled the close analysis of the different wavelengths of light emitted and absorbed by different substances.

In 1924, the next step was not the acquisition of more data or a new experiment. As we also observed in Chapter 2 in other cases, the moments of thought that set off the avalanche of new thinking were fired by the formulation of new *questions*. Like the perfect tool for a difficult technical task, the right question cracks open the closed husk of unprofitable ideas, and can release years of fruitful exploration of its consequences.

For Heisenberg, recuperating from acute hay fever on the pollen-free island of Heligoland in the early summer of 1925, the question presented in his mind was over which mathematical form might appropriately embed the act of observation into a theory of matter. Should a measurable quantity (such as position) be cast mathematically as an *operation* (so representing the carrying out of an observation) rather than a simple *quantity*? Put that way it seems rather technical and innocuous, but it initiated a landslide of change. Entities that were once thought objective (such as the position of a particle) became actions/operations, not objects, and within a few short months the puzzles of decades had become transparent windows onto a new understanding of the way the world works.

Heisenberg later wrote a significant work on the wider significance for thought of the new physics.[24] His own experience is not far below the surface when he writes

> *In the course of coming into contact with the empirical method, physicists have gradually learned how to pose a question properly. Now, proper questioning often means that one is more than half way towards solving the problem.*

This is so universally true in science that we often forget that it can come as a surprise to those who do not experience daily the search for the creative and powerful question, as one might search for a lost key to a locked casket. Perhaps we should not be so surprised; after all, most of the goals presented to school pupils constitute arriving at the right answers, rather than formulating the right questions. In science, as in any creative activity, the 'uphill' work is the formulation of the problem in the right way. Once this is done the finding of the solution is not the hardest work.

[24] W. Heisenberg, *Physics and Philosophy*. Transl. from *Physik und Philosophie*. London: Allen and Unwin, 1959.

To take another critical question from twentieth-century science as an example, and one with revolutionary consequences, we ought to recount the musing of Einstein's that he reported first occurred to him at the age of 16. He wondered simply what he would observe if he were able to travel alongside a beam of light (so travelling himself at the speed of light). He knew that light itself was an example of the physical interplay of electric and magnetic fields—thanks to the experiments of Faraday and others, and to the magisterial theoretical encapsulation of classical electromagnetism by Maxwell, all in the nineteenth century. He also knew that a consequence of the derived set of laws was that, once in free space, linked electric and magnetic fields followed the dynamics of a wave, oscillating and propagating at a single, well-defined speed—the speed of light. Although bewilderingly fast on a human scale (light could outpace Puck's wildest dreams, 'girdling the Earth' seven times a second), it is nonetheless finite.

'What would I see if I were to catch a light beam?' was the question that he formulated in his young mind. The answer, according to classical kinetics, was simple: 'I would see oscillating but stationary (relative to me) electric and magnetic fields'. But Einstein realised that such a phenomenon would *not* be a correct solution of Maxwell's equations, and would not be consistent with the experiments on which they are based. He could never catch a light beam. The innocuous, almost childish, question leads to extraordinary consequences—they begin with the realisation that one cannot, even in principle, travel at the speed of light. They rapidly move on to the dissolution of the Newtonian notion of universal time that flows equably for all observers, and of course lead eventually to the full *theory of relativity*.

It is perhaps because of the creative centrality of the question in all of science, both in the high-profile revolutions of relativity, quantum mechanics and statistical mechanics and in the day-to-day work of grappling with puzzling observations, that scientists are so impressed when they read the Lord's answer to Job. As we have emphasised many times before, this is not because they assume that the writer of Job was in any sense a scientist in our modern sense, with a set of developed methods and disciplines. But it does indicate the recognition of a foundational curiosity, observation and desire for understanding without which science is stillborn and powerless. It also recognises that powerful questions set programmes running—they are events of conception, of the birth of enquiry. Further, these reactions stem from an experience with the quality of questions. Not all questions are equally fruitful. Some are too far ahead of their time, launched into a vacuum of conceptual tools. Some are simply stillborn and without possible fruit. As formulated, seeking for 'storehouses of hail' is a question that cannot find an answer in its own terms. But planting the seeds of even ambitious questions—such as the explicit linking of laws of the stars and of the Earth—slowly opens up possibilities of thought and awareness to phenomena that prepare the ground for open scientific enquiry.

Love

That love plays a role in science will by now not come as a surprise. This is one of the advantages of following our project of locating science within a longer narrative of engagement with the natural world, itself supported within a theological story. Admittedly not a word usually invoked in the history and philosophy of science, those within its community of practice will recognise why we need to talk about the operation of love in the scientific project. By 'love' we mean here the super-rational (I do not want to write 'non-rational'), emotionally engaged delight in, and action to sustain the well-being of, another. A human project on the scale of science cannot work without its operation, though it may lie somewhat hidden beneath the surface of more obvious activity.

We have, likewise, not drawn explicit attention to this aspect of the developing relational theology with respect to creation that the Old and New Testaments contain. But love is not far below the surface in the biblical tradition either, wherever we look. The two Genesis creation narratives are (in this case, and unusually) a good place to start. God looks on his work at the completion of each day of creation from the third onwards, and 'saw that it was good'. Once we relieve ourselves of the freighted anxiety over literalness that clouds unnecessarily our view of this delightful text, we can appreciate the delight of a person fashioning a lovely world of light and life. The liturgy of days in chapter 1 repeats the affirmation of goodness in a climactic progression to the sixth and seventh days. After all is done God sees that 'it was very good', then on the Sabbath performs two significant acts: he 'blessed the seventh day and made it holy'. Blessing and sanctification are regal and priestly works of love. They endow fruitfulness within a framework of care.

The second chapter, as well as painting another parable of the creative act in different language, also transfers this delight and care for creation to humankind (Genesis 2:15).

> *The Lord God took the man and put him in the Garden of Eden to work it and to take care of it.*

The Old Testament scholar John Goldingay comments that, although the Genesis language is not explicitly covenantal, it essentially describes a covenant relationship between humans and nature. Recall that we did find explicit covenantal language hidden under one of the stones we overturned in our reading of Job. Eliphaz describes Job's future healing (once he repents) in just those terms:

> *At ruin and blight you will mock, and you will have no fear of the wild beasts.*
> *For you will be in covenant with the stones of the field, and the wild animals will be at peace with you.*

But a covenant is an expression of love, an unconditional commitment. It is the category with which we have traditionally understood marriage. Even the anguish of misunderstanding, disobedience and deception that wounds the created world in the events of Genesis 3 serves if anything to underscore the love for the world that is commanded of and engendered within humanity. The aspect of love that perseveres in spite of shortcomings, even 'irrationally' so, is shared in the painful engagement with the 'thorns and thistles' and the 'painful toil' of making the world fruitful.

To take one more of the Old Testament creation narratives we surveyed in Chapter 3, the agency of Wisdom as a little child in the enchanting eighth chapter of Proverbs adds colour and depth to the 'and it was good' of Genesis (Proverbs 8:30):

> *Then I was the craftsman at his side.*
> *I was filled with delight day after day, rejoicing always in his presence,*
> *Rejoicing in the whole world and delighting in mankind.*

It is this wisdom, enraptured by the created world, that the reader of Proverbs is commanded to find and to follow.

There are many links between the thematic categories that have served as grappling lines between our two stories of science and theology. In particular, love and pain share much in common; the endurance of pain is often the test and proof of love; a shared experience of pain can be a source for love. Childbirth sits at the focal point of both of these rays, so it is remarkable that this particular experience of pain is used repeatedly within biblical material tracing our relationship with nature. As Adam endures the thorns (arguably he was let off very lightly indeed), Eve receives the pain of childbirth as a consequence of their new self-knowledge. We recall the momentous passage in Paul's letter to the Romans in which he describes the current state of history as that of all creation 'groaning' in a metaphoric appeal to the same pains involved in bringing new life into being.

In a different context, Jesus also calls very explicitly on the metaphor of childbirth within his long 'priestly discourse' to the disciples before his death (John 16:19–28). The section deals with the pain of the disciples' lack of understanding (not about nature in this case of course, but rather about the approaching death of their beloved master) and the passage from current ignorance and confusion to future knowledge and the freedom that it releases. Jesus warns them, *'You will grieve, but your grief will turn to joy. A woman giving birth to a child has pain because her time is come; but when her baby is born she forgets the anguish because of the joy that a child is born into the world'*. In his map of the metaphor onto the disciple's experience, he deepens the meaning beyond simply turning grief into joy: *'though I have been speaking figuratively, a time is coming when I will no longer use this kind of language but will tell you plainly about the Father. In that day you will ask in my name'*. This most elevated example of 'pain in love' is about the passage from ignorance to understanding. Part of the new life that emerges is a greater ability to participate with the creator in the work of shaping the future. It is a glimpse into the way that Jesus thinks about the consequences of death and resurrection which, while not generally appearing in discussion of science and theology, more naturally does so within the much larger canvas on which we are setting the discussion.

Sadly, love itself can become diseased and twisted into power play on the one hand or debilitating need on the other. In terms of the natural world, the biblical metaphor for distorted love is idolatry. Both Proverbs and Jeremiah contain strong language that warns against an exploitative love of materiality. So the love of nature is not indiscriminate or unspecified. It delights but does not exploit, identifies worth but does not worship, puts up with pain for the sake of fruitfulness but does not seek to harm. Above all it is a love that seeks to understand.

Is there really any resonance of such aspects of love within science? A prevalent public view of the processes and methods of science sees cold logic, the application of a rulebook of experimental and theoretical practice and a disengaged, perhaps even dysfunctional, emotional approach by scientists to their subject. The influence of Karl Popper's *The Logic of Scientific Discovery* has cast a long shadow filled with these gloomy shades.[25] But his description of the 'scientific method' (it is intriguing to note that university science faculties do not deliver courses under this title) has frequently jarred to scientists themselves. Popper portrays us as working continuously to find evidence that refutes our current hypotheses, then when it does so, cleanly discarding them and proposing new ones. His balefully negative view arises, admittedly logically, from the general nature of scientific theory: our task is to deduce universals that apply to every conceivable specific instance. So Newton's law of gravity is a physical law because it applies to Jupiter, Saturn and to any planet orbiting any star, not to the Earth alone. Just one instance in which it failed would be enough to refute it as a universal principle. So for Popper one never proves a scientific law (he is right there, 'scientific proof' is another mythical beast in an imaginary conceptual zoo, on view in the cage next to the one for 'scientific method'). One may, however, and at one stroke, disprove it.

Even mainstream commentators on science have pointed out the severe shortcomings of Popper's analysis. Logical it may be, but the continual striving to prove one's own ideas incorrect finds no innate human energies on which to draw. Even at the level of scientific community, where rivalry might be tapped to drive the drama of refutation and counter-hypothesis, the real story is far more complex for two important reasons. First, Popper is silent on the provenance of the hypotheses themselves, yet without their coming into being there is no theory even to refute. Second, many subsequent critics have pointed out that even the heart of the action—the act of refutation of a hypothesis—is a far more slippery affair in practice than in principle. Confronted with new data that do not quite 'fit' a theory, it is usually far simpler to modify some of its assumptions, change the value of a parameter, build in a new effect, than to ditch the entire edifice. This is part of the insight of Thomas Kuhn, whose *Structure of Scientific Revolutions* in 1962 remapped our understanding of the history of science into 'normal'

[25] K. Popper, *The Logic of Scientific Discovery*. London: Routledge, 2002.

and 'paradigm-shifting' phases, when an accumulated mass of data eventually cause the old outworn intellectual scaffold to collapse entirely.[26] Neither Popper nor Kuhn identified the energies that provided the life support for theories and hypotheses struggling to come to terms with phenomena.

A more radical view of science was advanced later in the 1960s by Paul Feyerabend in his *Against Method*.[27] The title indicates the author's view of how well one can define a scientific method in principle: his mantra was 'anything goes' in science; he argued strongly for a social construction of science, and for a contingency of world-view on social setting. Few scientists would accept his more extreme relativist views, which would remove any anchors of knowledge into a real physical world, making the project an arbitrary craft answering to internal imagination alone. But the impact of *Against Method* has been immense. Equally few, as a result, would argue that science follows no social forces. But perhaps Feyerabend's most perceptive contribution is the insight he provides into the survival of theories and ideas, and to the essentially non-rational and 'counter-inductive' element that is not only admitted but actually necessary for science to prosper.

The key issue is the awkward nexus of conception and development of hypotheses at their very birth. No scientific theory is born antelope fashion, fully formed in limb and energy, able to run for itself and keep out of harm's way. Our ideas emerge far more frequently as a marsupial birth—inadequate, vulnerable and almost powerless. They are the conception of future hope, but require nurture in an uncritical environment before they are able to stand on their own two feet in the open field of academic survival of the fittest. In short they need to be loved into being. Feyerabend brought this to the community's attention in the starkest possible way: he appealed to the standard exemplar of all scientific revolutions, the replacement of the Ptolmaic and medieval geocentric cosmos by the Copernican heliocentric cosmology in the sixteenth and seventeenth centuries. Long portrayed as almost self-evident fact in the face of entrenched, conservative and religious blindness to the truth, the claims of the new world system were in actuality by no means self-evident.

[26] T. Kuhn, *The Structure of Scientific Revolutions*. Chicago: University of Chicago Press, 1962.

[27] P. Feyerabend, *Against Method*, 4th edn. New York: Verso Books, 2010.

It is worth examining this important story in a little detail. The great obstacle for Copernicus's theory might seem a little technical—it has to do with the variation in brightness of the inner planets Mercury and Venus—but it was enough to render the hypothesis demonstrably inconsistent with the facts throughout Copernicus's lifetime and into the telescopic age of Galileo. In the former model, all planets revolve around the Earth at an approximately constant distance, so would be expected to shine with similarly even brightness. But in the Copernican system Mercury and Venus are 'inferior' planets, orbiting the Sun more closely than the Earth does. This means that, although their separation from the Sun remains constant as they circle, their distance from the Earth varies considerably. Sometimes they reside on the opposite side of the Sun from an Earth-bound observer, at other times they are adjacent. Their apparent brightness should be much greater in the second relative position than the first. But anyone who has followed Venus as an evening star from first appearance to last (so tracking the planet's journey from the opposite to adjacent positions relative to Earth) knows that, while there is a mild change in perceived brightness, it is relatively minor—so minor, in fact, as to clearly refute (in Popperian terms) the heliocentric theory.

The resolution is found in the way that the planets shine—by light reflected and scattered from the Sun, rather than originating within the planets themselves. When close to us, most of the illuminated sunward face of Venus is hidden, the planet presenting us with its 'night side' and displaying a crescent phase resembling that of a new moon (and for the same reason). When, by contrast, Venus is far away on the other side of the Sun from Earth, we view almost its entire illuminated hemisphere. These two effects of proximity and illumination have opposite, and nearly cancelling, effects on the brightness of the inferior planets, rendering it almost as constant as would be predicted in a geocentric model. The rub, of course, is that the phases of Venus and Mercury are impossible to detect with the naked eye, and require a telescope to see. Copernicus died before even the most primitive telescopes were turned heavenwards and had no knowledge of the observations that would a half-century later disentangle the impenetrable knot that prevented his theory from universal acceptance.

What might come as a greater surprise is that Galileo, well aware of this problem for Copernicus, nevertheless praised him for promoting his theory in spite of the fact that it contradicted clear observational facts. The point here is that neither Copernicus nor Galileo was behaving 'unscientifically'. On the contrary, without a belief in the advantages —the simple beauty with which it explained the retrograde motion of Mars and Jupiter, the innate symmetry—of all that was admirable in the new world-view in spite of its inadequacies without persisting in placing their hope in those things in spite of the formal failings of the whole system, then the way to understand the solar system would have been closed to us. Radical new insights simply do not fall into place all at once. The process is messy, partial, even contradictory in places. The early modern astronomers simply placed their faith in the new idea, rightly believing that the wrinkles and inconsistencies would be ironed out in time. Above all, they cared, nurtured and defended their picture of the new cosmos as something they loved.

The need to cherish new scientific ideas irrationally does not arise only at the occasional moments of great scientific revolution. Michael Brooks in his book *The Secret Anarchy of Science*[28] charts the short cuts, the bluster of character, the use of scientific propaganda and more in the promotion of ideas that are short of experimental support. His case studies reinforce Feyerabend's claim that, in science, 'anything goes' where *Against Method* left off. Brooks does not condemn the behaviour he finds: 'I am filled with a new admiration for the anarchists of science. They make discoveries not despite their humanity, but precisely because of it'. But every scientist knows the innate conviction that an idea that is currently drawing contempt from all around contains, despite its fragility, the core of truth.

I came across a memorable example of the necessary committed love of a scientific idea early in my research experience. I was one of a number of physicists and chemical engineers seeking to explain the bizarre behaviour of elastic fluids composed of molten chain-like 'polymer' molecules: watched long enough these sticky fluids flow like treacle, but if deformed rapidly they bounce or snap like an elastic rubber.[29] Although a successful theory of single polymer molecules in isolation from each other had existed for some decades, attempts to extend this to cover the case of millions of overlapping polymers, twisted and tangled around each other, had simply failed over and again to explain several remarkable phenomena.

[28] M. Brooks, *The Secret Anarchy of Science*. London: Profile Books, 2012.

[29] It was my background in this topic that drew the attention of the peptide biochemists whose fluids had also undergone a transition from liquid to apparently elastic nature for no explicable reason; see Chapter 2.

The first challenge was the very high viscosity attained by these elastic fluids as the chain lengths increased. Second was the remarkable way that flow slowly set in over time under a steady load: there seemed to be a critical timescale before which the material 'bounced' and after which it flowed. Third, the fluids exhibited a strange 'thinning' phenomenon when the forces on the polymer melt were increased—the greater the force, the lower the viscosity seemed to become (in a familiar guise, this is a very useful property of paint that permits spreading by a brush while minimising running when subjected only to its own weight as a spread film). Finally, and the observation that most caught my attention, was the effect of grafted branches within the molecular structure of the polymers. Just one long side-branch per molecule was enough to alter the fluid properties by orders of magnitude—as if the topological form of the polymers were somehow more important even than their length (so a melted plastic composed of 'star-shaped' polymers could possess a viscosity millions of times higher than one composed of exactly the same polymers rearranged as linear chains). But an explanation for all these things seemed forever out of reach, buried in the mathematically unassailable complexity of billions of mutually entangled polymer molecules.

In the late 1970s a remarkably simple and beautiful conjecture was proposed by Sam Edwards, a theoretical physicist in Cambridge, and Masao Doi, at that time a visiting researcher from Tokyo. Extending an earlier idea applied to rubber networks by Edwards, and by the French physicist Pierre-Gilles de Gennes, Doi and Edwards suggested that we could think of any one of the polymer molecules as slithering within the entangled matrix of its neighbours. The surrounding chains would prevent any piece of the example chain from moving sideways very far, because the molecular chains are unable to pass through each other. However, any snake-like motion of the chain along its own contour would go unimpeded. Each molecule would therefore move, under the excitation of Brownian motion, as if it were confined to a tube-like region matching it like a sleeve along its entire length (the idea is illustrated in Figure 7.2). Only at the ends of the sleeve would the chain be free to diffuse unimpeded. De Gennes called this special motion 'reptation', reminded of a reptilian, snake-like, slithering. So all the unimaginable complexity of a picture that insisted on keeping track of all the chains at once was replaced at a stroke by an infinitely simpler one that focused on just one molecule at a time. All of the constraints and interactions with other molecules, and all *their* mutual interactions,

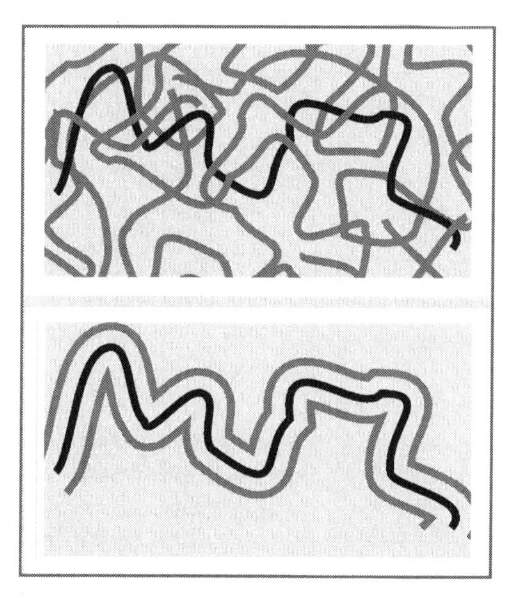

Figure 7.2 The idea of the 'tube model' of entangled polymers. The many tangles around a single chain (shown in grey to pick out its path) from its many neighbours are represented as a single confining tube, constraining its motion either to short excursions sideways or to random slithering motion along its own contour.

are simply replaced by the effective constraint of a tube. In pictorial terms, the 'many-body' picture sketched in the upper panel of Figure 7.2 is replaced by a 'single-body' model in the lower.

Professor Bill Graessley, a chemical engineer in the oil industry and later an academic at Princeton, New Jersey, had wrestled with the peculiar phenomena of polymer melts for decades. He described to me once the epiphany of the tube idea: 'It was as if a light had been switched on; everything made sense at once'.[30] The explanations did indeed tumble out of the idea in direct and simple ways. The timescale for allowing flow emerged as the waiting time for a chain to diffuse, by random forward and backward steps, out of its current tube into a new one, like a snake sloughing off its old skin. The longer the chain, the longer would

[30] A fuller account of the story and its later developments can be found in: T.C.B. McLeish and W.W. Graessley, The Doi–Edwards theory. In *Stealing the Gold: A Celebration of the Pioneering Physics of Sam Edwards*. Oxford: Clarendon Press, 2004.

be the time taken to escape a tube, in a calculable way.[31] Strong deformations tend to line up and stretch the tubes, but the chains retract rapidly within them, leading to the paint-like 'thinning' behaviour observed. And in one deft stroke of imagination, the critical role of molecular topology becomes clear: if but one long side-branch is added to each chain, then the tube structure is no longer a simple pipe, but itself must be branched, hugging the form of the polymer chain it contains. Reptation is immediately suppressed, all flow now relying on much slower retractions and re-expansions of the long chains now linked at a motionless branch point. Branched polymers trapped in tubes are like wriggling escape artists in straitjackets—it is not so simple for them to shed their more complex topological clothing.

No wonder the idea caught on, but—to my naïve surprise—not everyone found it beautiful, or even compelling. Perhaps, for some, the simplicity was affronting: polymer melts were just complex, and deserved something much more mathematically sophisticated. Others might have had a stake in alternative approaches, now threatened by the simple but powerful tube idea. Whatever the motivations, criticism from opponents was as strong, and at times as vitriolic as was the encouragement from its advocates. Not always recognised properly by its supporters at the time, but vital to understand, is that much of the criticism was valid: the predictions of the tube model were *not* all in line with experiment. The way that internal stresses were predicted to relax after a deformation, for example, was incorrect—the experiments revealed other timescales at work than the single 'reptation time' for escape from the tube. The predicted dependence of viscosity on molecular weight was close to experiment, but not within experimental error, and seriously wrong (the predicted viscosity was much too high) for shorter polymer chains. Anyone who wanted to find gaping holes in the theory was able to. Conference papers with titles such as *The Death of Reptation* appeared. Computer simulations failed to find the signature motions of reptating chains. Fundamental objections were raised against the self-consistency of the idea: if the tube around one chain was there because of other chains, then it ought to melt away as those other chains diffused, a process taking no longer for them than for the first chain.

[31] Simple physical arguments show that the escape time by curvilinear diffusion increases as the cube of the chain length—close, but not exactly following the experimental data.

Yet the idea hung on hopefully, or was rather nurtured and protected, perhaps irrationally by those who had fallen in love with it. Little by little the early anomalies were understood. The lower viscosity of shorter chain polymers emerged when the correct internal dynamics of the chains was added to the slithering reptation motion. Similarly, the search for the tube structures in such computer simulations was later successful, but only after huge increases in computing power made it possible to treat large enough systems. In the early years the lack of evidence from simulations, though actually due to lack of computing power, was often attributed to the misconception of the idea. But with new experiments other inconsistencies also emerged. At the time of writing, most of the community would accept that the tube model idea is fundamentally a good approximation to reality, and gets the key physics correct, but there are aspects of the true 'many-body' complexity of the problem that we still do not understand.

In a very small way, there is a shadow in this small revolution in a strand of material science, of the cosmological transformation of Copernican astronomy. Although perfectly possible to refute the idea in several ways using the data available at the time, an equally convincing argument ran that the idea was so compelling, and the insights and puzzles it had already solved so impressive, that remaining inconsistencies would be resolved as the theory became more sophisticated and new experiments available. So it proved to be. Copernicus was wrong about circular planetary orbits—they are elliptical, but the essential insight that the Sun is central to the solar system survived. The glaring problem of the brightness of Mercury and Venus vanished as soon as telescopes were turned on them and revealed that they showed phases like the Moon. Similarly the tubes of entanglement fields do indeed evaporate and reform, but on a range of timescales, most much slower than the escape time of a single chain. Other modes of motion also come into play, in a subdominant way to reptation, but essentially if experimental details are to be accounted for. Reality is indeed more complex than the simplest picture, but in a sense that more than tubes exist, not less. More is true of the solar system than that the Sun is central, but not less.

So it is that, with science great and small, taking delight in ideas that provide insight, the peering into nature with eyes that understand, is essential to giving them life and nurturing them to maturity. Those ideas build one by one into the project of reconstructing nature within our own understanding. As they do so they excite aesthetic responses within us: we find them beautiful, compelling, elegant. Sometimes we even come to realise that it is necessary to love them.

A Theology of Science: Participation in Reconciliation

We need to draw together some threads from our twin-track reading of wisdom and science. Through the lenses of our experience of science, on the one hand, and of the biblical theological story, on the other, we have traced how themes they both share have shaped and responded to these two deeply human endeavours. Long and linear history, the surprising human aptitude for re-imagining nature, the search for wisdom as well as knowledge, the ambiguity and experience of pain, the delicate balance of order and chaos, the centrality of the question and the questioning mind, and above all the experience of love are the lines that draw us to a larger narrative in which science can be framed.

Within all these themes the pattern of relationship has dogged us constantly. Science experiences the negotiation of a new relationship between human minds and the physical world. The nature language of the Bible is consistently employed to describe and develop the relationship of care and of understanding between humans and a world that is both our home and also potentially a frightening field of bewildering complexity. Although fraught with ambiguity, experiencing pain and joy in equal measure, knowing terror before the phenomenon of chaos as well as experiencing joy before its resplendent order, bewildered by ignorance yet granted hard-won understanding, the biblical theology of nature is consistently relational.

St Paul invested to a deeply personal degree in the nascent Christian communities with which he worked. None of these relationships was more turbulent than that with the small church in Corinth, to which he probably wrote at least three major epistles, two of which survive and are found within the New Testament. The pain of rejection at one point seems to have caused him a nervous breakdown, yet it is also within his dealings with this church that he composed one of the most sublime writings on love ever penned.[32] So it is perhaps not surprising that it is also within this correspondence that he rethinks the entire project of

[32] I refer to the passage often chosen for wedding readings from 1 Corinthians 13.

God's creation in relational terms, working around and towards the central idea of reconciliation. The argument begins with the fifth chapter of his second letter to the Corinthians, recapitulating briefly his picture of a 'groaning' creation, from the letter to the Romans, in longing for a more permanent form, which he calls *'clothed with our heavenly dwelling, so that what is mortal may be swallowed up by life'*. Arguing that those who have been baptised into the life with Christ can already view the world from the perspective of its future recreation, he writes (2 Corinthians 5:17)

Therefore, if anyone is in Christ – new creation;
The old has gone, the new has come!
All this is from God, who reconciled himself through Christ and gave us the ministry of reconciliation:
That God was reconciling the world to himself in Christ

The *ministry of reconciliation* is a stunningly brief encapsulation of the biblical story of the purpose to which God calls people. I do not know a better three-word definition of Christianity, and it does very well as an entry point for Old Testament temple-based Judaism as well. It acknowledges that there is work to do: relationships on all scales are damaged. Nation against nation, communities against communities, families, marriages, even the vital self-worth that describes people's relationship with themselves is often damaged. The biblical analysis of the reason for relational suffering is that an underpinning relationship—that between created humans and their creator—is itself in need of repair. Now is not the place to recapitulate the theology of incarnation, death and resurrection that explores the New Testament Christian *kerygma*, the 'good news' that a way to healing this broken relationship has been opened by God. Those wonders have been explained by others far better than I could.[33] But the Christian hope is just this—that mending that great relationship, voicing an answer to God's 'Adam, where are you?' call in the garden, and Jesus' agonised 'My God why have you forsaken me?' in his dying moments, releases the potential to heal so much more.

There is one relationship equally in need of healing that tends to be overlooked in expositions of Christian theology—perhaps humbler than the human ones listed above, but just as profound. It is the

[33] For a recent and radical account see N.T. Wright, *Surprised by Hope*. New York: HarperCollins, 2008.

relationship between humankind and nature itself. It is a relationship characterised by ignorance and fear in the past and to a great extent in the present also. It is often a damaging one: we exploit rather than care for the world, inadvertently tearing away vital components of the atmosphere, or replacing them with harmful ones; it can hurt us also with storm, earthquake and tsunami, with rising sea levels. But, like all hurt, it we do not need to shrug our shoulders and give up hope. As for all the human relationships, the consequence of the healing of the great relationship is that the covenant between humanity and the natural world can also be healed.

A theology of science, consistent with the stories we have told up to this point, situates our exploration of nature within that greater task. Science becomes, within a Christian theology, the grounded outworking of the 'ministry of reconciliation' between humankind and the world. Far from being a task that threatens to derail the narrative of salvation, it actually participates within it. Science is the name we now give to the deeply human, profoundly theological task of participating in the mending of our relationship with nature.

It is an extraordinary idea at first, especially if we have been used to negotiating ground between 'science' and 'religion'; as if there were a disputed frontier requiring some sort of disciplinary peacekeeping force to hold the line. It also makes little sense within a view of history that sees science as an exclusively modern and secular development, replacing outworn cultural practices of ignorance and dogmatic authoritarianism with 'scientific method' and evidence-based logic. But we have seen that neither of these assumptions stands up to disciplinary analysis on the one hand or to historical evaluation on the other.

Neither science nor theology can be authentic unless they can be universal. There can be no boundary delineating territory between them that does not immediately nullify their essences. We need a 'theology of science' because we need a theology of everything. If we fail, then we have a theology of nothing. Such a theology has to bear in mind the tension that the same is true for science—it has never worked to claim that science can speak of some, but not of other, topics. Science and theology are not complementary, they are not in combat, they are not just consistent—they are '*of* each other'. This is the first ingredient of a theology of science.

Just as there is no boundary to be drawn across the domain of subject, there is no boundary within time that demarks successive reigns of theology and science. It is just not possible to define a moment in the history of thought that marks a temporal boundary between the 'pre-scientific' and 'scientific'. The questioning longing to understand, to go beneath the superficies of the world in thought, to reconstruct the workings of the universe in our minds, is a cultural activity as old as any other. Furthermore, it is a human endeavour deeply and continually rooted in theological tradition. The conclusion is still surprising: far from being necessarily contradictory or threatening to a religious world-view in general, or to Christianity in particular, science turns out to be an intensely theological activity. When we do science, we participate in the healing work of the creator. When we understand a little more of nature, we take a step further in the reconciliation of a broken relationship.

Seen in this light, the global community and project of science in its many forms represents continuity of the biblical story that begins with creation itself, and of the making of humankind 'in the image of God'. It even sheds light on aspects of the centuries-long debate about what *in imago Dei* might mean. It explains why, for example, scholars have been confused over whether the subject of the first few verses of Job 28 refers to human or divine eyes, probing the very structure of the Earth from underneath.[34] That 'hymn to wisdom' concludes that God is wise because his perception reached to the ends of the universe, and calls wise humans to do the same.

It responds to the 'great commission' of Genesis and of Psalm 8 to 'rule over the works of [God's] hands'. But it does so in a freshly critical way, for a backdrop of participatory healing will not allow a crudely dominating interpretation of what 'rule over' might mean here. The framework of reconciliation is resonant with the participatory calling of the people of God throughout the Old and New Testaments. We have remarked in passing that Job, our ancient guide into the painful engagement with nature, is referred to by God both at the very start of the story and at its close as 'my servant' in fond terms. In later texts, especially in Isaiah, the servant motif is developed for Israel herself. The servant also becomes the messianic figure that will arise to heal and represent her in the renewing of the broken covenant relationship with God, and of the physical world itself. Isaiah even suggests that 'friendship' is as appropriate a category as 'servanthood' to describe our relationship

[34] See the discussion of Job 28 in Chapter 5.

with God in the face of the joint project of healing the world. Neither permits an exploitative dominance.

Participation in reconciliation is as strong a theme in the New Testament. Earlier in the Corinthian correspondence of Paul than the passage that we have briefly visited, he implores the Corinthians to see themselves as 'God's fellow-workers'. Perhaps most powerfully of all, the striking image invoked by Jesus in John 15 when he likens his relationship with the disciples to that between a vine and its branches works on many levels here. This vine casts a long shadow over any theology of discipleship: insisting that all of its ramifications are deeply participatory in the energy and person of God. It also returns to the nature imagery with which we are now familiar, and used to reading in both directions: here the Gospel writer borrows from nature to illustrate the Christian idea of living 'in Christ' in the ministry of reconciliation. But he is also alert to the long Hebrew tradition that the physical world participates in renewal just as much as the political, religious and personal. Finally, it draws on a running symbol of the vine as representative of the people of God; invoked by prophets in fruitless form when Israel fell short, and in stories of a tended vineyard to remind her of God's concern and delight.[35]

There is another surprise awaiting the explorer of this road to a theology of science. We noticed that among the 'clamour of voices' contesting the relationship between theology and science was the accusation that scientists today behave like the 'priests of a former age'. This is meant of course in a pejorative sense: the image projected is the priest as representative of oppressive authority, of the exclusive keeper of mysteries, of an agent whose power-play is to keep a population ignorant and fearful. It is not often commented that such a collection of associations is actually of a corrupted idea of priesthood, and that there are other models of this historical religious role which bring quite different values to the word 'priest' which have been overshadowed. The Old Testament notion of priesthood was in many ways the perfect opposite of the dark stereotype we sketched above. The priests of the temple were the personifications of the task of reconciliation, not the obstacles to it. Theirs was the charge of the festivals that reminded the covenant people of their calling, and of the sacrifices that earthed into an effectively timeless religious praxis the foundational acts of reconciliation initiated

[35] See, for example, Isaiah 5:1–7.

by God. They were the agents of a reconciled relationship of people with God. The New Testament transformation of priesthood is actually an amplification of a healthy notion of priest as enabler, rather than a negation of it.

The theology of priesthood has also been coloured by the conflicts of reformation and counter-reformation. It is hard to see clearly through some of the smoke of this battlefield, especially when attempting to work out what the apostle Peter was trying to say when he articulated what a Christian Jew might do with the notion of priesthood (1 Peter 2:9):

> *You are royal priests, a holy nation, God's very own possession. As a result, you can show others the goodness of God, for he called you out of the darkness into his wonderful light.*

The notion of a 'universal priesthood' in New Testament theology has generally been taken in the sense of mission. Believers are called to mediate between God and humankind by sharing the gospel. But it is a weakened sense of priesthood, for it loses all differentiation in calling. A relational approach to a theology of science can call on a much stronger, yet still transformed, idea of priesthood. In this view, we are all priests because we are all the mediators of reconciliation in our own special domains of life and work. Scientists are priests in the sense that they are specifically at work in transforming the broken relationship between people and the physical world. Special work is required because the task calls on long training and particular skills, but this is not a priesthood that then prevents the new understanding over ignorance, new delight over fear, new nurture over exploitation, from benefiting others. On the contrary their work is on behalf of the people and requires regular participative celebration.

Does such a theology of science do meaningful work for us? Does it provide any avenues to resolve the painful cross-currents around science in society? Does it suggest new tasks? These must be the test for any endeavour of this kind. It has taken long enough to climb the hill from which we might see science within a Judaeo-Christian world-view, and outline a theology *of* science that begins to circumvent a relentless contest over ground between theology *and* science. The next chapter will begin to explore what consequences a theology of science within a 'ministry of reconciliation' might lead to.

8

Mending Our Ways, Sharing Our Science and Figuring the Future

We have worked our way through some dense undergrowth along the forest trails of 'love of wisdom of natural things'. In the journey we have kept close company with theological themes, principally from the Bible, some equally thick with thorns and briars. It has proved rather natural to travel with this narrative partnership—and it has taken us towards a new perspective on science within a long human tradition. Surprisingly, in spite of the jarring edge of the word 'science', and of a shrill public debate fuelled by the assumption of an irreconcilable conflict between world-views carrying the flags of 'science' and 'theology' (or 'faith' or 'belief'), we find that an enquiring engagement with the natural world has a recognisable and central place in Judaeo-Christian biblical narrative. Some close parallel readings, of 'science stories', on the one hand, and biblical wisdom, on the other, especially the book of Job, have helped us to identify where a perception of conflict arises. Public debate runs into trouble initially from falsely projected notions of what 'science' and 'theology' actually mean in practice. These distorted images are then doubly conflicted with a falsely imposed 'geometry' of science and theology within the space of ideas, assuming an underlying relational and bounded structure inconsistent with both of them.

Paradoxically, the problem is not that proponents of the 'conflict' metaphor have overstated the case, so much as they have not gone far enough. For both science and theology claim not only to be able to speak about some things that the other also does, but each, by its nature, demands to speak about everything. Their scope is universal or it is nothing. In particular, we must not avoid the conclusion that each needs to speak authoritatively about the other. Science and theology need to be 'of' each other. Science, as Daniel Dennett has pointed out, needs to analyse the anthropological, sociological and neurological basis of religious thought and practice. A 'science of theology' is as natural as it promises to be helpful—it is a doubly misplaced reaction of religious people to run or hide from a reflective and scientifically informed critique of the belief and praxis of faith traditions.

But science is blind to purpose—it has no 'teleological' methodology or goals. It shines light on what happens in the universe and how it happens, it digs deeply down below the surface of phenomena into underlying interactions, causes, symmetries, connections between apparently disparate phenomena, emergence of complex structures from underlying rules, but it cannot articulate stories with goals and values, however much it illuminates the social and mental systems that support those stories. Science is also very needy—the very resources that it depends on require a publically understood narrative of its purpose and goals, its risks and promise, how it arises from and contributes to being human, why and how our communities should value it.

Theology must also speak about science, for it also must talk about everything. It does so by recognising that the universe contains persons, communities, relationships, living experienced stories. It defines and interprets the narrative structure around them. It works with the radical, but not ridiculous, idea that personhood might extend even beyond the confines of physical nature. It risks the hypothesis that the universe is shot-through by information, by 'word', because of such a person. Because theology observes and construes stories, it is able to discuss purposes and values—it can speak of, and ground, 'teleology'. So it can and must talk about the purpose of science within its story and its values. A theological story might find science irrelevant; it might find it reprehensible or dangerous, or affirming or wonderful. But in any case it needs to have the conversation within a wholehearted grasp of science as a human endeavour.

We have identified a strong, central narrative thrust of the biblical tradition of the human—the mandate for *healing or reconciliation of broken relationships* (as argued, for example, by St Paul[1]). The last chapter explored how the scientific process might function within such a theology, rather than as counter to it. This is not a common approach to debates on the relationship of science and theology, but it is arguably more historically faithful than either confrontational or complementary viewpoints. It is

[1] Memorably in the second letter to the Corinthians, a correspondence itself shot-through with the themes of pain and reconciliation (2 Corinthians 5:18,19).

also logically faithful—in this way of telling the story, neither science nor theology are epistemologically 'penned in' but are both free to talk about everything under the Sun. In particular they can talk freely about each other. It is certainly a suggestive starting point for an ethical discussion on scientific research, but, as we will see in this chapter, opens doorways onto much more. For the roots we have traced situate science as that part of the wider, and more ancient, task of healing that enables the reconciliation of humankind with the *physical world* (in a parallel way to the better-trodden ethical grounds of reconciled individual psychologies, marriages, families, communities and nations). As well as *relational*, this theological framework is resolutely *participative*—we are God's co-agents in the process of bringing nature into a right relationship with its creator and with its human inhabitants. The vision replaces a broken relationship of ignorant exploitation of the world, on the one hand, and the threat of re-emergent chaos, on the other, with a renewed husbandry of creation, marked by a shared wisdom and understanding. Science becomes part of *shalom*, of achieving peaceful and sustainable co-existence with our world. Far from being 'irreconcilable' with religion, it becomes one of the great holy tasks of a people themselves created out of love, and intended to grow in wisdom and understanding.

It is the purpose of this final chapter to take out our map once more, but now, rather than navigating some overgrown pathways of the past, to look instead towards the uncharted territory in front of us. If we are seriously suggesting that the geometries by which the relation of science and religion have been understood in the past are deeply flawed, from irreconcilable conflict on the one hand to peaceful and complementary irrelevance[2] on the other, then this must have consequences for how we think and act. It must suggest new ways of framing science within our communities in education, politics and the media, and in our churches and beyond. The much closer embrace of science and theology that our story suggests, such that each admits to being a legitimate property and object of the other—that they are naturally 'of' each other, means also that there will be fresh ways of understanding

[2] The Cambridge theologian Sarah Coakley has criticised the approach of 'naïve complementarity' to science and theology, which leaves no one perturbed from their initial position, in her recent Gifford Lectures *Sacrifice Regained: Evolution, Cooperation, and God* in 2012. Coakley draws on her collaboration with Harvard evolutionary theorist Martin Nowak to approach teleology in science from another direction. See <http://www.abdn.ac.uk/gifford/about/>.

evolving religions tradition in the century before us. Put another way, the question before us is this: from the mountaintop view of nature we gained from the book of Job, what are the consequences of our new perspective for the paths we choose to take down into the territory on the other side?

The first consequence is unexpected, but intriguing. For from this elevated viewpoint other trails appear more clearly, and travellers with guidebooks other than the Bible turn out to be not as far away from us as we might have thought.

The Witness of Other Traditions

We have traced the outlines of a relational theology of science through just one tradition (Christian, with its Old Testament Jewish background). Such a narrow focus might legitimately be challenged on the grounds of its limited cultural scope, although it has certainly allowed us to explore deep and rich veins of tradition that are arguably the dominant cultural underpinnings of Western thought. However, a remarkably similar pattern shows signs of emerging from contemporary Jewish, Islamic, Vedic or classical viewpoints. While it is not in the scope or the purpose of this book to do justice to these traditions at the level at which we treated the biblical material, it is worth noting that the trailheads of 'nature walks' within these other traditions do not look impossibly at odds with our Judaeo-Christian findings, for they also speak of mutual comprehensiveness and reconciliation between human and material. More strongly, we might suggest that a comprehensive theology of science is itself a strong motivation and resource for tackling interfaith dialogue from a new perspective. If faith traditions have any current challenges in common, surely among the foremost would be an urgent need to think through the place of science in their world-views. If this can be done in communication and collaboration, then dialogue between the faiths would be deepened at the same time as gaining a needful understanding of the place science holds within them.

We may even start with a conversation with atheism. From a perspective of science stories within comparative world-views, a dialogue between theistic and atheistic world-views is just one subset of a broader class of conversations. Yet here too both ancient and modern resonances with the idea of participatory reconciliation are strong. A fragment from Socrates' contemporary, the playwright Euripides, contains the benediction: 'Blessed is he who learns how to engage in inquiry, with no impulse to harm his countrymen or to pursue wrongful actions, but perceives order of immortal and ageless nature, how it is structured'. This classical tradition has an inheritance in the secular thought of our own time. We recalled George Steiner's remarks in his hauntingly relational and personal assessment of language and meaning[3] 'Only art can go some way towards making accessible, towards waking into some measure of communicability, the sheer inhuman otherness of matter . . .'. Of course a scientist's reading of this claim can feel like a slap in the face (only art?), but it serves to help us evaluate whether or not our research really approaches at some level a restored communication with the 'otherness of matter'.

Without such a resonance of Platonic, Aristotelian and Pythagorean classical thought with the Hebraic tradition we have examined, the eleventh- and twelfth-century European scientific renaissance would not have received the impetus from Christian scholars that it did. Anselm, Grosseteste, Bacon and others were happy to adopt a scientific method from Aristotle of 'resolution and composition' (we would today map these terms inexactly onto 'deduction' and 'induction') within a clear mandate to explore nature's workings. They found it natural to embed such early scientific thinking into a theologically informed world-view. As we saw in Chapter 2, this process generated interesting challenges to their received classical wisdom as well, such as the physical cosmogony of Grosseteste's *De luce*. Contemporary statements that inherit the ancient Euripidean blessedness of the ability to perceive order and structure in nature are usually couched not in this classical language, nor in the biblical language of wisdom, but in terms of explanatory power. Steven Weinberg,[4] for example, writes

> *We have been steadily moving toward a satisfying picture of the world. We hope that in the future we will have achieved an understanding of all the regularities that we see in nature, based on a few simple principles, laws of nature, from which all other regularities can be deduced.*

Although strongly resistant to any theistic interpretation of the world, the motivations of hope and its partial and eventual satisfaction are

[3] George Steiner, *Real Presences*. London: Faber, 1989.

[4] S. Weinberg, Can science explain everything? Anything? In *The Best American Science Writing* (ed. M. Ridley). New York: HarperCollins, 2002.

articulated here in a way that any scientist would recognise. 'Hope', 'satisfaction', 'understanding', 'regularity' are also strongly resonant with the theological story of movement from divorce to reconciliation, from ignorance to knowledge and wisdom. The trailhead that starts from commonality of purpose (in this case in regard to understanding nature) between theism and atheism points to more distant possibilities. A 'theology of atheism' has a great deal of material to work with from the very wisdom literature we have mined for stories of nature. The book of Ecclesiastes, for example, looks deep into the meaninglessness of a purpose-free cosmos (Ecclesiastes 1:13)

> *I applied my mind to study and to explore by wisdom all that is done under the heavens. What a heavy burden God has laid on mankind! I have seen all the things that are done under the sun; all of them are meaningless, a chasing after the wind.*

Compare this striking passage from the Old Testament with atheist science writer Peter Atkins–'we should be aware that deep down we, like everything, are driven by purposeless decay: that is why we have to eat.'[5] The voice of the 'teacher' in the book of Ecclesiastes reminds us of Job at his most weary, and also that just because you admit to there being a God it does not automatically bring with it the reward of meaning. He agrees with Atkins' view in our own time that the study of nature can as easily lead to the chasm of nihilism as to a narrative of purpose. Science, if nothing else, enriches the dialogue between atheism and belief, but this is *not* because it produces evidence for one world-view over the other. Rather, science raises for believer and unbeliever alike the question of the human relation to the material, and how meaning might arise out of it.

In modern Judaism, we find a remarkable account of science developed by the American Orthodox Rabbi J.B. Soloveitchik, an influential scientist, philosopher and theologian for much of the twentieth century. In his book *Halakhic Man*[6] he attempted to outline the contemporary consequences of living by Talmudic law. Among its interconnected lines of thought, Soloveitchik's discussion of 'scientific man' develops the idea that science, within Jewish law and practice, calls principally on the creative capacity of humankind:

[5] Peter Atkins, *On Being: A Scientist's Exploration of the Great Questions of Existence*. Oxford: Oxford University Press, 2011.

[6] J.B. Soloveitchik, *Halakhic Man*. Philadelphia, PA: Jewish Publication Society of America, 1984.

> *This notion of hiddush, of creative interpretation, is not limited solely to the theoretical domain but extends as well into the practical domain, into the real world. The most fervent desire of halakhic man is to behold the replenishment of the deficiency in creation, when the real world will conform to the ideal world and the most exalted and glorious of creations, the ideal Halakhah, will be actualized in its midst. The dream of creation is the central idea in the halakhic consciousness—the idea of the importance of man as a partner of the Almighty in the act of creation, man as creator of worlds.*

In Judaism there is no 'fall from grace' of humankind or creation as there is in Christianity. The Adam and Eve story in Genesis bears a different interpretation—that in regard to the human, people have always had the potential to disobey, and that in regard to the material there is 'deficiency in creation' in which we participate, as creative partners with God, to make good. A sense of 'finishing the work of creation' runs closely alongside our narrative of 'reconciling with creation', a deeply creative programme in itself. The linking idea is that only human agency could recognise anything like a 'deficiency' in creation, for as far as we know only humans have the capacity to imagine an alternative order of relationship in the universe to the one that actually exists.

Any notion of deficiency or replenishment of what is missing in the created order must lie, once more, in the relational gap between the human and non-human. This alternative perspective from Judaism underscores just how ambitious the goals of our scientific programme and our capacity for transformation really are. Part of the 'deficiency' in Soloveitchik's thinking is the current inability of man to assume a harmonious role as a partner in creation. We do not currently enjoy the 'satisfaction' of Weinberg's explanatory knowledge, so are incapable of wisely applying what we do know. Chief Rabbi Jonathan Sacks points out that Judaism has a prayer of blessing and thanksgiving for scientists, whether they are believers or not, because what they do is understood to be within the compass of God's work.[7] The continuity of Old and New Testaments is stronger here than their discontinuity. A New Testament Christology might respond to such a Talmudic vision of replenishment by identifying the source and sign of the hope humankind must invest in its realisation, by pointing to the resurrection. So it might add to the Jewish vision, but it certainly does not detract from it.

[7] J. Sacks, *The Great Partnership: God, Science and the Search for Meaning*. London: Hodder and Stoughton, 2011.

The relationship of the Islamic world to science is complex, many layered, more fraught with tension even than that of Christianity. This is a potentially explosive element within the politically charged relationship of the Abrahamic faith communities and the nations in which they dominate in the twenty-first century. Yet, as we have noted in our brief encounter with medieval developments in our 'stories of science', the early twelfth- and thirteenth-century flourishing of scientific thinking in Europe (and, by direct descent, the later flowering of the 'Enlightenment' in the seventeenth and eighteenth centuries) owes a huge debt to the Islamic culture of the early Middle Ages. Mechanics, medicine, optics, astronomy and the development of the philosophy of science from Aristotle all made huge strides in the Islamic world of the ninth to thirteenth centuries.

Muslim scholars from this extensive period made outstanding and influential contributions to science. Equally important to progress was their development of open scientific debate, argument and disagreement. Al-Kindi's (801–873) discussions of Euclidian optics are known to have influenced Grosseteste in the early thirteenth century. Avicenna (980–1037) wrote extensively on practical medicine as well as an account of experimental method and an effective critique of ancient astronomy. Averroes (1126–1198) commented extensively on Aristotle's *Physics*, developed early notions of inertia and conceived what we might now term a complementary approach to (Aristotelian) science and Islam. An anxious debate over the reasons that the later medieval world of Arabic Islam did not realise the promise of those early centuries in the development of science, but rather gave way to the Renaissance European tradition that it had itself furnished intellectually, cannot be discussed with any attempt at adequacy here. It may be that Islamic theology is more tightly constrained to a more 'received' interpretation of the Koran than Christian or Jewish theology allows itself with respect to the Bible, or that the religious authority structures of the twelfth and thirteenth centuries in the Islamic world did not permit the freedom of thought, and of interpretation of its sources, in the same way that characterised the Christian West at that time. But where more self-critical Muslim thought does appear, then it demonstrates strong resonances with the participative and creative narrative we have discovered.

So, for example, the twentieth-century Islamic philosopher Muhammad Iqbal[8] urges that strict interpretations of Koranic sayings on God's action in

[8] M. Iqbal, *The Reconstruction of Islamic Thought in Islam*. Oxford: Oxford University Press, 1934.

the world, which have traditionally been taken as pre-determinative of human action, and proscriptive of human freedom, should rather be thought of as a realisation of potential within a created universe, a 'free creative movement', and as an invitation to human sharing in the task of creation. This notion takes us down to the deepest roots, where we find that Islam, Christianity and Judaism share their two most fundamental stories. The first is of the significance of creation at the beginning, and of the universal interest of the creating God in all of the cosmos. The second is the particular story around the significance of human beings in double relationship with God and the world, exhibited and developed originally in the character of Abraham. Exploring the nature of the unfolding of those stories in time, and of the development of the human–material relationship, is certainly a fruitful way of understanding how science serves human culture. Exploring them together across the Abrahamic traditions may also point towards a badly needed road to understanding each other.

Glancing across from our 'nature trail' through Judaeo-Christian wisdom towards other paths through the forest, and to the travellers on them, suggests the first, surprising and perhaps somewhat hopeful consequence of a relational and participative theology of science. When deep engagement and mutual understanding is ever more urgently needed between the monotheistic faiths, and when their universal demolition in the face of a purely atheistic scientific worldview is not a credible future, a project that draws on deep common narratives of creation and healing might prove a source of understanding between religions themselves. To greater or lesser extent (Judaism is the least troubled) all religions have a problem engaging with science. At the very least this appears as a lack of confidence, at the worst an explicit antagonism. Our journey suggests strongly that this is because they have forgotten or ignored the strands of their own traditions that would embrace science as the modern (and postmodern) manifestation of a deep narrative within the purposes of those traditions.

Yet rather than celebrate and enjoy science as the empowering of an ancient mandate to light up the world and care for it, religious communities all too often perceive it as a threat and so fail to provide the cultural and theological context it actually needs. They also miss out on the benefits of a scientific perspective on faith itself; some are even terrified at the prospect. But if it is true that science belongs inside, rather than outside, the long narrative of relationship healing with the natural world, itself part of the purpose of faith traditions, then there is nothing to fear from the mutual comprehensive embrace of science and religion that we have termed as being 'of each other'.

Drawing on the relational and participative strands of our theological and scientific journeys, our next task is to ask what the consequences might be for a healthy contemporary and future practice of science with human societies. Let us try to sketch a few roads of departure from our current position. In each case there will be two signposts: one from each of the paths we have followed in the foregoing chapters. From our stories of science within its cultural context we have already become alert to stresses, confusions and miscommunications that themselves point to a need for a richer narrative of purpose for science, a clearer shaping of its course within our communities. From our theological track we have collected by now a rich set of ideas that might connect with these needs. We will venture a short way along avenues that the science stories and theological tradition seem to point out. Some of the places they lead to will challenge and surprise us.

A Long Journey: the Timescales of Science

A theological perspective on science immediately recalibrates the proper length and pace of its story. The natural timescale for a research project, one that plays even a small part in mending our currently broken relationship with creation, must be a long one. Deep wounds heal slowly, and require more than superficial restructuring of tissue. Or, to take a theological view, the double-edge of eschatology (it implies as much for how we live now as for what may happen in a distant future) requires both incisive action and patient waiting. Even though there is indeed reason to celebrate sudden, even decisive, entry of hope into the world now, this does not imply its immediate full realisation. The creation of the world was a rich and extended process, taking aeons of time unimaginable in duration; there is no reason to expect that a programme of participative restoration will be rapid.

The old wisdom literature often repeats the theme of the long and difficult search for wisdom itself, and one aspect of the painful theodicy that so often accompanies biblical contemplation of the physical world is that the 're-search' itself is to be a long and painful one. Think again of the thorns and briars of the journey out of Eden, the puzzlement of the psalmist confronting the glory of the heavens with the world out of joint, of Job's indefatigable yearning for vindication, of the universal groaning in St Paul.

We will have more to say on what a juxtaposition of the ideas of 'contemplation' and 'science research' might mean later, but already we can see that our groundwork throws the actual practices of scientific research into sharp relief. Without the long perspective of the wisdom tradition one can become so accustomed to the modus operandi of the social and political machine that regulates science today that any sort of ethical critique becomes at best groundless and at worst blind. The first shock that meets eyes now turned to contemporary research practice is the contrast of our anticipated long timescale of science with the relentless political urgency that fuels the international research engine. In his valedictory 'Festival overture' lecture to the Edinburgh Festival in 1996, George Steiner typifies the 'noonday' cultural activity of science: 'Theorems will be solved, crucial experiments performed, discoveries made next week and/or the week thereafter'.[9] He means to contrast a healthy scholarly community (ironically allowing his audience to think of science as one such) with a diseased one (that, according to Steiner, of the humanities). But to anyone within the scientific research community the remark is chilling. Increasingly, the pace of discovery is determined by a painfully yoked pair: a political lack of trust with a political demand to deliver. This in turn drives the invention of increasingly complex and weighty tools of accountability and performance management.

True of no country more so than the UK, three decades of 'research assessment exercises' in this country have cost hundreds of millions of pounds and millennia of person-years of work. Every 5 or 6 years, every university researcher and every department is measured in detail for the quality of their output and its impact. Such colossal accounting has reshaped our landscape of learning. Research activity and output is linked to funding, which is itself linked to assessment scores, which are linked to research activity and output. . . . The analogy offered by former Archbishop of Canterbury Rowan Williams[10] with the storyteller Scheherazade in *Tales of the Arabian Nights* is exactly right. Every day new theorems will be proved because new stories must be told. The cliff-hanging

[9] G. Steiner, *The University Festival Lecture: 'A Festival Overture'*. Edinburgh: University of Edinburgh, 1996.

[10] R. Williams, Faith in the university. In *Values in Higher Education* (eds S. Robinson and C. Katulushi). Cardiff: Aureus Publishing, 2005.

hiatus of today's story becomes the grant proposal that fuels the tale of tomorrow. If the next novelty is not forthcoming the axe may well fall on the research group.

An added dimension of concern at this issue of 'science as storytelling for survival' is its palpable appearance at several levels. As individual researchers compete for the ear of the sultan, so do their universities. Even at a national level, the biennial (in the case of the UK Treasury) 'comprehensive spending reviews' demand stories of economic promise from the entire research community (represented in the UK by the research councils). It is all too easy to equate success at these survival games with the maximising of financial return. There is dwindling political pressure to support research projects that run long, slow and deep. Evaluation is, in the end, measured in economic benefit. Other values that reflect the deeper goal of a participative reconciliation with nature do not feature in any current audit.

The politically driven pace of science has consequences also for its wider influence and perception. To take one example, the current critique of science as overly functional, and separated from 'human' values and aspirations, is by no means new. Famous cries against its assault on human and spiritual are present, as we saw in Chapter 1, in Blake, in Keats, in Dickens, in Flaubert. The core of these complaints contains a voiced fear that science will actually destroy the means of our reconciliation with nature (Keats' accusation in *Lamia* of 'unweaving the rainbow') rather than nourish it. Jacques Barzun made the memorable analysis in his thoughtful critique of the developing value structure of science that 'science is not with us an object of contemplation',[11] an observation from the 1960s that is no less relevant now. It is still the case that research in universities is a surer route to professional preferment than the more contemplative activities of scholarship and teaching. I know of a few wonderful exceptions of scientists who are prepared to spend weeks working on pedagogical material that lets deep ideas connect, relationships assume their proper perspective and draws selflessly on the best work of others, without any immediate 'research results' in view, but they are very rare.

Particularly when access and participation in higher education, and the necessary preparation for it, are under reassessment, a rediscovery of both the pace and the depth implied by a contemplative research and

[11] Jacques Barzun, *Science the Glorious Entertainment*. New York: Harper and Row, 1964.

teaching methodology is a discussion that needs to be reintroduced into the public arena. There are wonderful practical examples of embodying the long-term and gentler vision of a 'natural philosophy' over short-term 'science'. The aptly named *New Horizons* mission to Pluto was happily reinstated, following budget cuts, in response to a public petition organised by a single high school pupil. NASA then chose an unusually young team to constitute the mission scientists in the preparation to launch in 2006, simply because the arrival at its remote destination of even this very speedy craft would not be before 2015. I find the implicit value placed on the continuity of the team admirable. At a much more individual level of decision to invest in the long term, the mathematician Andrew Wiles quietly resisted the pressure to publish regularly, devoting himself throughout the 1980s and 1990s to finding a proof of Fermat's celebrated last theorem. This task would be measured by the yardstick of a decade rather than a year or a month. It would have been difficult for him, if not impossible, to have achieved this in a UK university (he was supported during this period at Princeton University, in the USA).

Data suggesting that we are currently entering a period of very rapid climate change have reignited a healthy debate about our responsibilities to later generations, rather than simply to ourselves, in terms of our planet's sustainability. It is manifestly part of the wonder of being human that we can think like this of a future beyond our own lifespans, with distant horizons informing our investment of resources and time, both at individual and at political level. But we also need to recognise that long-term thinking and values will be under threat if that story goes quiet. The closely connected elements of the long biblical story we have surveyed: human responsibility to cherish the world, to recognise that we need to work with nature in knowledge and wisdom rather than ignorance, that the future does not lie in a return to some classical idyllic past but in transformation, that there are questions that must be asked but that will go unanswered for a very long time; these constitute a foundation for a healthy reappraisal of timescales.

A Global Community of Practice

Shared values drive communities, so what might we look for in a reconciliatory scientific community? A natural consequence of a relational context for scientific ethics would be a self-sustaining communication and debate among a 'community of practice'. The biblical threads we have followed cover the entire tapestry of a human community in partnership, working with and within the created natural order. So such a community would be defined by this purpose (rather than by national or other political boundaries) and would maintain a continual reflection on its future direction and past account. The existence of a scientific community of practice, tasked with informing and changing our relationship with the physical world, is by no means an obviously reasonable notion in a 'post-modern' world of individuals and private readings, especially as it would be one in which individual achievement, though celebrated in the context of the whole task, would not be emphasised or accorded levels of prestige that obscure the project's underlying goal.

There are reasons to be optimistic, at least at first sight, about the nature of a research community suggested by a relational agenda for science. The integral activity of the research conference, for example, where a single research area is regularly discussed by the international practitioners who lead it, sustains social as well as professional relationships, welcomes young members, suggests new directions and celebrates the spoken word in debate and discussion. As such it has become an emblem of an underrated strength of academic research, namely the ability to build and sustain truly global communities that transcend cultural and linguistic boundaries. Furthermore, as Polanyi pointed out long ago, these research communities are enabled to function because they do not impose on themselves the intrusive accountability we noted is increasingly demanded from outside, but instead recognise the essential currency of trust. This is why cases of 'scientific fraud' are so mercifully rare (but, in consequence, of such high profile).

Yet there are reasons to suspect that the best of this practice is in danger of erosion. The last 20 years have seen the publication of an unprecedented number of nationally commissioned reports on unethical practice in science.[12] Some of the pressures that arise from the increasing pace of publicly funded science and that lie behind these voiced concerns we have already discussed. But there are others that are increasingly placing scientists in positions of conflicts of interest from which it

[12] For example, Committee on the Conduct of Science, *On Being a Scientist*, 2nd edn. Washington, DC: National Academy Press, 1994; D. Fanelli, How many scientists fabricate and falsify research? A systematic review and meta-analysis of survey data, *PLoS ONE* 2009, **4** (5), e5738.

is hard to escape, and that could induce fissures into the fragile nature of healthy research communities. A heightened value placed on competitiveness generated by new systems of accountability, and research funding is an inescapable badge of honour. Scientists will find themselves at one level urged to play for the prestige of their university, but simultaneously and at another level for a national research network. In the case of the European Union, funding channelled through the commissions in Brussels constitutes a third, higher, level at which suddenly one's rivals must become colleagues over and against other large global players. Moving in the other direction, micro-managed funding models within institutions set up artificial rivalries between departments in the same university.

In this increasingly complex environment of externally urged enmity it becomes harder to sustain a vision of the globally situated yet locally cognate communities that embody a reconciliatory and contemplative research agenda. A researcher embarking on a career in the centre of the maelstrom of counter-currents is beset with questions. With whom, and at what stage of development, should I share this idea or question? At what level of association do I declare a 'conflict of interest' in assessing or refereeing a publication or grant application? These activities in particular are at the core of the system of 'peer review' that has evolved as a formalisation of self-reflection within the scientific community. As currently practised, peer review does represent a real safeguard against obviously flawed or unnecessarily repeated work, but its anonymity and lack of access to original data in most cases (to name but two limiting, if necessary, features) mean that it is itself not immune to abuse. Scientific papers infamously flawed with fabricated data, since withdrawn, passed unproblematically through peer review. Every scientist has experienced the grinding frustration of a grant proposal turned down on the basis of an ignorant (willingly or otherwise) report from an opponent held out of reach of argumentation by the peer review process.

So, just as equivocal peer review is placed under increasing pressure, some of its inherent drawbacks become more evident. In particular, there is an urgent need to address how to avoid suppressing the surprising, the innovative, even the revolutionary research programme under the double weight of accountability on the one hand and engineered rivalry on the other. Panels and committees tasked with ranking over a hundred 10-page research proposals before teatime are unlikely to deal effectively with the very few that would fundamentally change or expand our vision of an area of biology or physics, let alone bring interdisciplinary patterns of thought together in new ways, however reliably they may compare incremental research comfortably within established programmes. In the UK, some tentative moves are being made at the national level to fund individuals and groups with promising records without a detailed dissection of their proposals. Yet even these, when they are not opposed by appeals to equality of opportunity, rely at approval stage on the solid and reliable, yet unimaginative, peer review process.

A community overloaded with these highly formulated evaluative processes in setting the research agenda is also tasked with policing the results of research programmes. So in the area of publication a similar set of pressures arises. Financial and personal credit comes to those who publish not just in quantity, but in those journals accorded with 'high impact factor'. Recent high-profile cases of series of papers in the high-impact journals *Science* and *Nature* have shown how a web of dysfunctional motivations and relationships within a research community can result in the publication of extensive accounts containing fabricated data. Large research teams working at speed and under pressure from their director begin to economise on discussion and mutual checking of results. The journals themselves require a steady stream of novel and apparently cutting-edge research to fill their pages and emblazon their decorative front covers. Furthermore, the high-profile end of the scientific publishing spectrum has become manifestly *fashionable* in its selection of material and authors.

The potential instabilities inherent in such implicit collusion are obvious, especially when the governing pace of a research area has been accelerated beyond a respect for reflection, repetition, sifting and criticism. By degrees, attention is drawn away from the *contemplative substance* of the science, from establishing a firmer and deeper reconstruction of the physical world, to the epiphenomena of visibility and prestige of the scientists themselves, and the publications that channel their results to the public. Remarkably, the pressure to publish a result before rivals do differs considerably within sub-fields of science. A physicist colleague, used to an extended period of investigation of a problem before even beginning discussion of a possible publication, was startled to find that the biologists in the same institution felt constrained to be thinking always of 'the next publication' rather than 'the next problem', lest their rivals publish on it first. These Scheherazades seem not to need the fear of a vicious sultan to spin their tales—they keep them coming anyway under a self-imposed fear of tomorrow.

The continual pressure to tell a new story, any story, has another instantly recognisable effect on publication: that of the multiplication of secondary or derivative results in great quantity. Even in the very narrow sub-fields of research typical of a scientist working today, it is impossible to read the majority of the mountainous publication stream whose titles clamour their relevance. A professor, now retired, on leafing through the many pages of yet another new journal in his field muttered the response as frequently thought as it is rarely spoken: 'there simply can't be this many clever people in the world'! The scientist might therefore be puzzled to read a heartfelt criticism of precisely this phenomenon, but aimed exclusively at the fields of the arts and humanities, by one of their own scholars, in deference to the 'primary' perceived content of the sciences. Going further than in the 'Festival Overture' we quoted in the last section, George Steiner, in *Real Presences*, imagines a 'primary city' in which all writing is fundamentally creative art, and no secondary, derivative or parasitic forms are admitted. The way to talk about a poem is to write another poem. Crucially the proliferation of the secondary in literature is related for Steiner to the breaking of the 'contract of meaning' between words and world. One of the attributes of post-modern studies in literature—the centrality and self-justification of the text—becomes for him a force for the debasement of texts themselves in a proliferation of self-referential commentary. Steiner sets this movement, as we saw that he does the general pessimism of his field in the 'Festival Overture', against his perception of relative academic health in the sciences.

But the scientific grass on the other side of Steiner's fence is not really as green as all that when examined close-up. The same pattern of proliferation in written output is hauntingly connected to a scientific version of Steiner's 'broken contract'—for when science is superficial it, too, fails to connect with its object, and now the malady is palpably worse than its literary version. It is arguably academically legitimate in a department of English to write texts discussing texts. But a discourse among a scientific community that has lost an essential physical referent has also become a worthless heaping up of words. This is not meant to be a sweeping characterisation of the current scientific literature, but we may well be concerned when the average readership of a scientific paper published today numbers no more than a handful.

We seem to be confronted by a hedge of thorny problems that threatens the health of a global science community of practice. Growing public and political mistrust of science, an increasing pressure to paint as glowing a picture as possible of scientific findings, a tendency towards fashionable topics, an intensification of a hierarchy of prestige attached to where, rather than what, scientists publish, the increasing difficulty of newcomers to the community in becoming accepted, the confusing lines of competition for advantage—all of these have a common source of misplaced value.

There are some surprising lines suggested by situating science in a theological context that speak directly to this need to rebuild value within the science community. They are surprising because they echo one of the 'clamorous voices' that we heard in Chapter 1 accusing scientists of becoming the 'priests of our age'. The thought was driven, of course, by images of a corrupt priesthood in an exploitative and repressive power relation with its community. But what happens to that metaphor if we replace the concept with a more biblical priesthood narrative?

This is the story that begins with the identification of Old Testament priests as servants, rather than overseers of the people, committed to a particular reconciliatory task (the observance of the Temple, itself symbolic of the point of meeting between people and God). Priests are privileged only in the sense that they have a particular and specialist task to perform on behalf of the rest of the people. It is a visible and accountable role, and one in which the community also plays a responsive part. Like many theological themes its biblical history converges to a focal point in the New Testament with the commission of Jesus as the 'great high priest' in the ultimate work of healing relationships, then diverges again into the community of the new people of God as St Peter's 'priesthood of all believers'. As we saw in the last chapter, if this means anything, then it means that every person has a reconciliatory role, but that these are exercised in diversity of callings.

In this light, our relational theology of science has a somewhat astonishing consequence: scientists do indeed constitute, in this sense, a priesthood, but with no different status from the priesthood of factory workers, chefs, teachers, builders or carers. They simply have a special domain of healing and nurturing work to do, and this is on behalf of the rest of the community to whom they are accountable. In particular their work should not be closed in on itself, but requires a developed aspect of service to others. A twenty-first-century community of practice that contemplates the model of priesthood in rethinking its future development sounds wildly anachronistic, but in the light of the threats facing science in public life, on the one hand, and our radical reappraisal of the human narrative history of science, on the other, is one of considerable force. Far from 'guarding access to the secrets on which our life is based',[13] a priesthood of scientists carries a manifest responsibility to provide access to the public good they create, and to adopt a servant rather than celebrity culture. This in turn drives us to think about how we might rethink public participation in science.

A Shared Science of Relationship and the Task of the Church

What do we see when we look at science practitioners from the perspective of the wider community, now through the lens of a long human story of healing broken relationships? The first consequence is that the goals of science become much more widely shared than we currently perceive them to be. They would not be limited to the small group of people that is charged with tasks requiring long training. There would be, as we saw in the last section, no elitist 'priesthood of scientists', but instead an understanding that theirs is a servant community engaged in a publicly shared project. Those with special expert roles are therefore also charged with listening to and communicating with others more widely. The community would support a continuous traffic of communication, and in particular an exchange not purely of the knowledge that comes from research, but also of the wisdom that comes through participation in the scientific process, and in the understanding that it engenders. Wisdom would also come from the very act of sharing with the wider community.

Tentative but promising moves in this direction have been made since the 1990s in several countries that support science from public funds. Research councils in the UK now support academic scientists as 'media fellows'. PhD students have been funded to work within high schools as young role models for pupils considering science. US grant funding bodies insist increasingly that 'outreach' plans accompany research grant proposals. Significant media profiles have been developed by scientists who seek to communicate a love of doing science, its depth and

[13] Angela Tilby, *Science and the Soul*. London: SPCK, 1992.

unique ability to engender wonder and its human rewards. But if our long narrative analysis carries weight, then these constitute no more than the first steps in a radical engagement of a wider public in the scientific project. For if the goal is a mental, emotional and practical reconciliation with a strange, and estranged, material world, then this cannot be achieved vicariously by a few on behalf of many.

Some thinkers we have already met seem to have glimpsed the possibilities: Jacques Barzun's 'object of contemplation' and George Steiner's 'performative science' look at music, art, literature and theatre and wonder why science too cannot establish equivalent bridges of mutually enjoyed communication between the dedicated performers of the art and the public which supports them. In some aspects progress here will be rediscovery. Michael Faraday's Friday evening discourses at the Royal Institution in the London of the 1830s and 1840s were 'theatre' in every sense—they would be sought after by the same audience which might be found on the following evening at the Theatre Royal, and wearing the same formal dress. At one discourse in 1846, an audience numbering over twice today's recognised safe capacity (450) squeezed themselves into the elegant tiered lecture theatre to hear Faraday describe experiments connecting magnetism with the polarisation of light. Journalists complained that they were unable to enter to report the event.[14]

It is worth exploring a little further the narrative context of science in that earlier period we visited briefly once before in Chapter 2 where Faraday was extolling the virtues of Robert Brown's investigations into microscopic motion. Part of the backdrop to Faraday's scientific world was the role science played in the Romantic age of the eighteenth century.[15] The period saw the Herschels combining a love of both astronomy and music, Davy writing poems to show his friend the poet Samuel Coleridge, who replied with results of his own experiments. It was the era which saw the stunning use of light and shade by the painter Joseph Wright of Derby in his depiction of scientific apparatus, transfixing the attention of adults and children alike. It is surely not coincidental that the retreat from such full-blooded cultural appropriation of science began as, from the mid-nineteenth century onwards, science

[14] And so additionally providing confirmation that in 1848, as now, journalists would turn up at the last minute.

[15] Beautifully captured in Richard Holmes' *The Age of Wonder: How the Romantic Generation Discovered the Beauty and Terror of Science*. London: Harper, 2009.

established its culture of fragmented sub-disciplinary expertise. Natural philosophy gave way to physics, chemistry, biology—and to their associated societies of experts. The initial change was a narrative shift: from natural philosophy as a romantic engagement with nature full of exhilaration and terror to a story of knowledge accumulation by 'scientists' whose methodologies grew ever beyond the grasp of anyone without the long training available only to their elite.

The Romantic narrative is, we should note, not the same as the story of participative reconciliation that emerges from the biblical tradition. To be sure it shares with it the elements of wonder and energising curiosity, the exhilaration of the astonishing human ability to reimagine the workings of nature within our minds. But it has none of the controls and none of the purposeful humility which come from a commission to work with the creator in Genesis or Job. Nor does it recognise the prioritisation of wisdom over knowledge as embedded in the narratives of Proverbs or the prophets. So this is also the age that gave birth to Mary Shelley's novel *Frankenstein*, telling the chilling tale of the monster that a human dared to create but then refused to love and nurture. No other story has infiltrated the narrative world, and freezing into place there so expanded the cracks between the community of science on the one hand and its public contemplation, grasp and enjoyment on the other.[16] A red-top newspaper title editor faced with a science story two centuries after Shelley's first edition and with a mind to stir up controversy, only has to reach for the prefix 'Franken-' to make the point. Subsequent imagery (and careless journalistic usage) has largely forgotten that it was not the creation of a conscious being in the first place that drew Mary Shelley's censure but the refusal of Dr Frankenstein to enter into relationship with the creature thereafter, to live (and love) with the consequences of his act of transformation in nature.

Could a new narrative, emergent from our older biblical strands, serve to reinvigorate a deeper participative engagement in science that builds on the best legacies of the Romantic era, while avoiding its uncontrolled tendency towards nemesis and the will to power without responsibility? Such a public narrative would retain the shared energy of engagement with nature, its reillumination and husbandry of the Romantic narrative, but would bring alongside a love of wisdom, a relational and responsible

[16] Reviews in, for example, Jon Turney's *Frankenstein's Footsteps*, 2nd edn. New Haven: Yale University Press, 2000.

mandate of care, and a clearer vision of purpose within a framework of service. There is every reason to build hopefully towards such a reappraisal of science in public life, but there are very serious challenges to meet first in the three spheres of education, media and the churches. We explore them briefly in turn.

The educational challenge, in at least the UK context, is entangled with the relentless drive towards a focus on 'core' disciplines. Science and 'faith' are often perceived as 'incompatible' in surveys of young people in full-time education. But this conclusion is unlikely to be a result of a thought-through epistemology when the curriculum tells them so without a word spoken or written. In a survey of teenagers in UK schools in 1994,[17] 21% of boys agreed that 'science has disproved religion'. Interestingly this was true of only 11% of girls; and among boys but not girls, this opinion correlated strongly with those who also agreed with a strongly 'scientistic' *narrative* of how we know things about the world. So just because we do not present or inform pupils with a clearly articulated cultural history of science does not mean that they grow up free of narratives—they are more than capable of picking up the implicit ones that we leave lying around. This is not to lay blame at the feet of our school system, which works under strong constraints of a national curriculum and a dearth of material and resources for any seriously multidisciplinary engagement, let alone one as complex as the human and cultural role of science. The faith–science nexus is an especially challenging topic for schools. But there is a very urgent need to equip the next generations of adults to think about science and its consequences for being human, and our global future, if they are to take their responsibility as citizens constructively.

The 'long-timescale' aspect of our story needs to enter strongly here too, for the artificial conflict narrative, which is implied by the way we currently frame science in schools, stems in part from the polarisation between sciences and humanities in general. In my own experience as a guest speaker in the excellent but declining sixth-form 'general studies' courses, I always ask those pupils who are not studying any science courses the reason for their choices. Those sitting further back, with whom I have perhaps needed to work a little harder to win their

[17] W.K. Kay, Male and female conceptualisations of science at the interface with religious education. In *Christian Theology and Religious Education* (eds J. Astley and L.J. Francis). London: SPCK, 1995.

attention, might volunteer that they thought the science options 'too hard'. But those (typically) at the front, who have been engaging deeply and bright-eyed with the discussion, and are clearly able enough to excel at anything they choose, will typically complain that sciences did not seem to offer them the space for imagination or the avenue for creativity that they were looking for. A sword pierces my heart at this point, for reasons that will be clear from the 'stories of science' we have followed. For without creative imagination of the highest order science suffocates. The soaring imagination required to conjure up the questions of a contemporary Job, to conceive of tomorrow's theories with the power to unlock today's vast unsolved mysteries, must equal that of any poetry, music or art.

Such able thinkers and creators are the very people who in a generation's time will need to understand why and how we do science, what it might be for, and how it must be employed in shaping their successors' future. One simple way to begin planting the skills they will require would be to enrich selected topics within a science course with an investigation into its history of discovery. Even a glimpse at the characters involved, the blind alleys, the incomplete data, the partially formulated questions and the extraordinary moments of inspired conception when the fog lifts a little would correct the impression of so many that science is simply a 'received' body of knowledge. Nor would it be possible to retain the impression that scientific knowledge can be accrued through the routine application of some 'method' free of all imaginative content. Without some knowledge of the cultural history of science it is impossible to begin an informed approach to the specific engagement of science and religious belief. But if we can dispel a current set of misinformed myths, such as the denial of any pre-Enlightenment science worth the name, we will be the more able to equip the newest members of our adult communities with the understanding they so desperately need.

What is true of education is equally true of the media, for all the rich potential for life-long learning that they make possible. And if communication within the scientific community is already exhibiting the fault lines that point to a failing grasp of a healthy relational ethic, it is perhaps not surprising that issues of public ownership, communication and trust in the scientific process and its depiction in the media are currently under serious scrutiny. In the UK, enquiries such as those of the House of Lords report on the public understanding of science, commentaries exemplified by Onora O'Neill's 2002 Reith lectures[18] and the regular business of bodies such as the Advisory Committee on Novel Foods and Processes have identified a 'climate of suspicion' surrounding the process of science. A singular case in point, fuelled into a frenzy of public narratives of conspiracy, was the so-called 'climate-gate' leaking of emails between some UK climate scientists.[19] The furore reflected in part a perceived lack of independence of publically funded research from government, but it had much more to do with a complete failure to grasp how science is actually done, the doubts and uncertainties that accompany even the smallest steps.

When combined with the powerful prejudice of climate-change denial the counter-currents created a whirlpool of misunderstanding. But, as Jon Turney has pointed out,[20] it would be wrong to complain superficially of an 'anti-science' attitude that sets out to contain and to distance itself from scientists. More consonant with our working model of the fundamentally shared and public domain natural to science is, once more, the analysis of Angela Tilby[21] (and Simone Weil before her in the case of technology-driven markets) that scientists run the risk of becoming a new priesthood (in an exclusive, not a servant, sense), 'seeming to guard the key to knowledge, to have access to transcendental truths which the rest of us could never hope to understand'. If these cries come from educated and intelligent people who feel shut out from a world of rich insight disappointed that the scientific community seems to offer no way in to a shared reading of its process or results beyond the superficial or patronising, then part of the problem must lie within the media that channels the debate and projects the image of the science community. As we have already seen, the theological resonances are very strong, and

[18] O. O'Neill, *A Question of Trust: The BBC Reith Lectures 2002*, Cambridge: Cambridge University Press, 2002. <http://www.bbc.co.uk/radio4/reith2002/>.

[19] For US and UK reports on this incident, see Todd J. Zinser, *Examination of issues related to internet posting of emails from Climatic Research Unit*, 18 February 2011. Washington, DC: Office of the Inspector General of the United States Department of Commerce. <http://www.oig.doc.gov/OIGPublications/2011.02.18-IG-to-Inhofe.pdf>; Secretary of State for Energy and Climate Change, *Government Response to the House of Commons Science and Technology 8th Report of Session 2009–10: The Disclosure of Climate Data from the Climatic Research Unit at the University of East Anglia*. London: The Stationery Office, 2010. <http://www.official-documents.gov.uk/document/cm79/7934/7934.pdf>.

[20] Jon Turney, *Frankenstein's Footsteps*. New Haven: Yale University Press, 2000.

[21] Angela Tilby, *Science and the Soul*. London: SPCK, 1992.

a lesson from the church history of corrupt priesthood is surely of the latent social instability and impoverishment of such a perceived combination of power and exclusivity.

Our media is very mixed in its contributions to these and other public images of science, both negative and positive. This is even true of high-quality broadcasting channels such as the BBC's Radio 4, which exemplifies models from a 'court-jester' approach in interviews with scientists on its news and current affairs programmes to the personal, playful yet respectful and engaging style of programmes such as *Material World*. A recent broadcast of Radio 4's current affairs programme 'Today' for example, concluded as usual with two items of wider cultural significance: the first on a recently discovered cosmological structure of immense proportions, formed by exploding galaxies, the second a reflection on the significance of a prominent twentieth-century French novelist. It was remarkable that the scientists were reprimanded for using the 'difficult language' of the term, 'a simplifying assumption' on air, while the literary debate received no such censure for its extremely polysyllabic discussion of philosophical and novelistic influences in nineteenth- and twentieth-century French literature. Even the media aimed at the most educated public, at peak listening time for influence, is embarrassed by its inability to connect in a confident way with the cultural life of science.

The theology of science that we have begun to explore suggests, however, two specific opportunities for change. The first springs from the difference between external pressures on the media and those on education. In particular the continual race for ratings drives editorial policy in ways hugely damaging to any hope of progress beyond the 'clamour of voices' in our ears with which we began. The problem is that conflict is *entertaining*, much more so than a strait-laced imparting of new scientific knowledge, but it is false to assume that these are the only alternatives, requiring that the conflict narrative be stoked by creating false oppositions. The opportunity here is the discovery of other dimensions of entertainment in the story of science that derive legitimately from 'tensions of opposition'.

The natural vehicle that emerges from our story is the tension of light and darkness, of ignorance and understanding. One reason that the projection of science in the media is less faithful to its passions and pursuit within the science community than is the case for the arts is the focus on end results, on knowledge gained, rather than the paths taken to get there, the imaginative leaps and years of hard work, or on the humbling topic of what we do not know. Eroding the exclusive aspects of the 'priesthood culture' of science without losing the special roles of experienced experts, a deeper and more faithful media engagement with science would invite an aspect of audience participation that in the arts is a respected and essential part of the creative process. It would at the same time find ways of being more seriously critical yet celebratory—an experiment worth making, both for its wider human enjoyment and in supplying the vital requirement of a political community that understands the process of enquiry behind technical risk and uncertainty.

A responsible community of practice and an enlightened media of communication are even together not enough—to complete a healthy engagement with science we need receptive public institutions. To create a vision of a developing relationship of wisdom with the natural world needs leadership from within communities—if this is embedded theologically into the continuing story that began in the Bible, then this leadership is a clear opportunity for the church itself. At present much is wanting. At the national level, it is very rare that senior church leaders comment on science, and, when it happens, with any confidence. Locally within congregations, scientists become bewildered at the uncertainty with which their vocations are valued. A young person leaving their home church for theological college will receive public attention, an interview at the front of church and very possibly an invitation to write newsletters home. But the student leaving the same church for a degree course in physics does not have to be told in so many words to know that their choice is much less valued—the muted farewells and absence of any public recognition of their important step say it all too clearly.

Church teaching programmes, through sermons or study courses, are almost uniformly silent on science. And, as in parliament, the tiny proportion of church leadership with a science background only reinforces the distance, a profoundly unsafe one, between the church community and our engagement with nature.

But there are signs that when fear is replaced by confidence, when a vision of the theological context for science is grasped and when imagination and energy combine there is real ground for hope. A recent 'closed door' discussion of the possibilities of new genetic therapies was arranged by bishops and other senior church leaders with a group of foremost UK scientists in the field. Participants from both backgrounds were pleasantly surprised at the level of engagement of the other, the thoughtful openness that characterised the meeting and the emerging consensus of goals. In particular, the way that churches draw wide sections of the community together around (among many other things) a reflective agenda of what we *ought* to be doing with the technologies and political powers at our disposal means that they can be a vital resource in national debate on sensitive technological issues.

From a very different perspective, the underplayed 'contemplative' possibilities for science can also find an effective public channel through the embedded nature of churches within their communities. A church in Bristol, regularly visited throughout the day and maintaining an open-door policy, decided to mount an interactive exhibition of thought-provoking science. One exhibit was a beautifully machined 'double pendulum'—the beguilingly simple system whose complex and chaotic behaviour we encountered in Chapter 2. Visitors were able to set the pendulum going from many different starting conditions and gaze, often in wonder, and the subsequent crazy gyrations of the device. Local scientists also provided material for the exhibitions, emphasising the often unrecognised fact that scientific ideas are being created in any city containing a university, research centre or research-based industry. The emphasis was on personal engagement with observations and questions, together with the beginnings of ways of explaining, but without requiring a neatly packaged 'expert' answer to every one. Of particular note for our explorations in this book was the rather natural avoidance of the direct discussion of 'conflict' between science and religion. This was simply a contemplative installation within a space and building that symbolised enquiry, engagement with question and truth in a natural way.

I had a related experience in my home city of York in 2012. The city decided to celebrate its broad tradition of science and technology through nearly 100 sponsored weather-proof poster displays on the walls of public buildings. Walking guides could be downloaded from a website, which also contained further information in text and graphics on each example. Colour pictures were of very high quality, whether of specifically stained protein markers in cells from the university's biology department or the contours of vibrational modes on trumpets from a local musical instrument manufacturer. Each image represented a new or a deeper way of 'seeing' nature. Many were stunningly beautiful.

York's own inhabitants and many visitors clearly enjoyed the idea of rooting science into their own community.

A reporter from BBC Radio York spent several hours with me (on a very quiet and therefore necessarily *very* early morning) visiting and talking through a personal selection of the posters. We must have recorded about 90 minutes of material. I was not sure how and when it would be broadcast; interestingly, the station found it natural to do so over a number of weeks within the context of its Sunday morning programme which, though maintaining a broad perspective, retains a background theme of the city's churches and finds it natural to include contemporary songs of worship with choices from the current Top 40. We did, I admit, conclude with a discussion of not only the possibility but also the necessity of an engagement of science with religion that avoided the superficial traps of the tired conflict narrative—it was an opportunity not to be missed. But the overwhelming majority of the material was reflecting on the science or technology and its significance, or on simple ways of understanding it. The thoughtfulness in the reporter's approach meant that the gentle religious context seemed to the editors to be the most conducive one. Feedback from listeners was in agreement.

There is an aspect of the message about science, broadcast both internally and externally by the church, that does need a stronger, clearer and more courageous voice before this vision of a church reconciled with and celebrating science can be realised. Of greater currency in the USA than in the UK, though by no means silent there either, is the so-called 'fundamentalist' doctrine of a young (6000 years or so) Earth and a 'literal' interpretation of the first (but not therefore of the second) chapter of Genesis. Developed further, it insists that all species were created fully formed by special fiat of God, and are not connected by any evolutionary tree, demonising Darwinian evolution and therefore most of biological science in the process. I have not in this book marshalled any full-frontal critique of this largely sociological phenomenon (it is not in continuity with any mainstream tradition in Christian theology, but emergent in the late nineteenth and especially the twentieth centuries), for when the biblical material is surveyed in its entirety, and allowed to inform a theology of nature authoritatively, the shallowness and distorted reading of a '6-day' exegesis becomes evident. Paradoxically, while promoted typically by those who make high claims about the 'authority of scripture', it betrays a shallow disrespect for the richness of the Bible. It is an aspect of what Rowan Williams, former Archbishop of Canterbury, terms a 'faithless kind of faith'.[22] Falling into a trap for the ill-informed that Augustine warned about a millennium and a half ago, it invites the (in this case well founded) accusation that the church is irrelevant. It paralyses the Christian message of reconciliation and hope, plays into the hands of authoritarian power structures, and imprisons the minds of the thousands of honest believers, especially when children, into dark cells of ignorance, when a God-given ocean of truth awaits enjoyment just outside.

The church has a word for this sort of wrong teaching, the sort that makes it harder for those who hear it to participate in the 'good news' of the Christian gospel, but instead to become confused and fearful—it is called a 'heresy'. The word is not so current these days—it has of course uncomfortable historical associations—but there is no progress to be made by any form of peaceful coexistence with a wilful denial of our path to reconciliation with the natural world.

Once we can close off such false culs-de-sac, and understand what the biblical voices are saying, alongside our experience of doing science, we can take the journey signposted by the writers of Genesis, Psalms, Job, Romans, and embrace the work of scientists within the same grasp as the other work of the church, just as the purview of science becomes at last the same landscape as that of theology. A participatory and reconciliatory theology of science has a very practical consequence for the worship and teaching within churches. It also suggests that science is not only compatible with, but an integral part of, their mission. There is surely opportunity for wider experimentation along the lines of these examples alone, and much more besides, confined only by the limits of imagination and energy.

'Pure', 'Applied' and the Healing of a Fragmented Academy

I talked in a single sweep of 'science and technology' in the examples of public engagement above. Of course the two have often appeared in the same breath, both today and in former ages, as we have noticed in both

[22] R. Williams, Faith in the university. In *Values in Higher Education* (eds S. Robinson and C. Katulushi). Cardiff: Aureus Publishing, 2005.

the conflicted political framing of science policy in our own times and the moving account by Gregory of Nyssa on the interaction of air and water at the deathbed of his sister Macrina. We need, however, to tease out the interlocking but distinct tasks of science and technology within our theological framework. Even if we begin the journey with a clear notion that 'lighting up the world' (science) and 'changing the world' (technology) are very different things, our understanding of their places of meeting is bound to shift and extend in the light of it. Deeply connected, for example, to the pace and pressure on science that we have already noted is a political framework that prioritises the instrumental utility of science over its other human goods.

Here our theological path takes us to perhaps a surprising place, for it does not deny the utility of our reconciliation with nature. It is not so much that we find a lack of value in 'knowledge for its own sake' in the wisdom tradition; it is more that such an idea is simply not recognised—all knowledge connects. The double biblical strands of 'knowledge' and 'wisdom' combine a reciprocal and practical ability to work with natural materials and processes which is indispensable to an enlightened understanding of nature itself, as well as flowing from it. We found celebration in the knowledge of seasons and plant life intimately linked to making the soil fruitful by farming (Isaiah), the knowledge of the underground structure of earth and rock wedded to the very technology of mining from which it comes (Job).

One could say that biblical wisdom introduces us to the idea that science needs technology to give it eyes just as technology needs science to give it wings. We had better re-examine our simple notions of the division of disciplines into the 'pure' and 'applied' categories so entrenched in (at the very least) the Anglo-American academic world.[23]

As we have found when teasing out other consequences of our narrative, the pressing issues for science and technology start in the undergrowth, rather than the top canopy, of the forest of our institutions. So we should return to one of the places where research ideas take root: the university campus. What we find is a landscape of knowledge rather less integrated, and admixed with arguably less wisdom, than we might expect from the intellectual inheritors of Job 28 or Proverbs 8. In any

[23] The much wider connotations of the European languages' translations of 'science' (*Wissenschaft*, *la science*, *la scienza*, . . .) as knowledge in general represent echoes of a broader and more connected view of knowledge in German, French and Italian tradition.

case it is a much more fragmented one. For the fissures in our framework of knowledge that drive notions that 'pure' and 'applied' might apply to entire disciplines are found far wider than simply at the meeting point of science and engineering. They criss-cross the near two-century-wide territory that opened up between the humanities and the sciences themselves as the nineteenth-century programme of specialisation spread. They create, for example, fiercely contested boundaries demarking the 'human' and the 'physical' within the discipline of geography, and arm frontier posts between the 'quantitative' and the 'qualitative' social sciences.

Such insistence on a divided disciplinary world denies the common and reciprocal good, and the mutual begetting of knowledge and wisdom. Most fundamentally, it dislodges the cornerstone of our theology of science—that learning is also about the healing of relationships. A theology of 'science' (in the Anglo-American sense of the physical and natural sciences) must become in the end a theology of 'Science' (in the European tradition of all knowledge, of nature and the humanities). In the current jargon of the academic world, it is radically *interdisciplinary*. That is to say, while recognising that knowledge must be acquired through its categories (think again of the structure of the Lord's answer in Job, with its careful structuring of strophes on weather, sky, inanimate and animate nature), completing the process of *understanding* the world can only come through a synthesis of these domains of knowledge. This is itself of course a task for communities, not for individuals.

Wilson Poon and I have explored the interdisciplinary implications of the theological and ethical path we are attempting to map elsewhere,[24] but harnessing a motivation for healing the academy is not enough; we also need to rediscover the threads that bind disciplines together, however divorced they may seem to have become. The arts and humanities, and the sciences too, share some of the maladies of George Steiner's critiques (of inwardness, the valuing of secondary and tertiary rather than primary learning, of a trivialising and relativising of knowledge)—they all experience and suffer the pejorative use of the word 'academic'. Yet because they all set out on a publicly owned project claiming to be central to our well-being, both our arts and sciences have more in common than is often admitted. Returning once more to Steiner's haunting call

[24] T.C.B. McLeish and W.C.K. Poon, How many cultures? 'Real presences' and the healing of the academy, *Interdisciplinary Science Reviews* 2001, **26**, 167–72.

to 'go some way towards making accessible, towards waking into some measure of communicability, the sheer inhuman otherness of matter', we can now identify in his words the task of reconciliation with natural matter (for which 'communicability' is surely a prerequisite) to which all disciplines are urged. We are not the first to have unearthed a suggestion that a more holistic approach to knowledge and learning is desirable; what is lacking is more practical—realistic ideas on how to achieve this, rather than reasons to wish for it.

Fortunately, the theological narrative does more work for us beyond identifying once more the chasm that Steiner and others urge us to cross as an ancient human mandate, and one in which all types of learning have a part to play—it also suggests where to look for connections between those currently splintered disciplines. Suppose we bring together a number of our story's strands: the invitation to engage with questions, to reimagine the cosmos beneath its surface, the insistence on wisdom's deep perception, and the universal experience of winning of knowledge through pain. Together they present us with the shared experience of *creativity* with *constraint*.

The idea of 'constrained imagination' is an experience that both arts and sciences recognise. It lies at a deeply structural level within both when we look closely and without prejudice. Again, terminology creates an obstruction when use of empty phrases such as 'the exact sciences' and 'scientific proof' serve to convince creative people that there is no room for imagination in the scientific process. Without imagination there can be no hypotheses, no possible accounts of 'subterranean' structure of nature to evaluate by the constraints of observation and artful experiment. Karl Popper recognised in his *Logic of Scientific Discovery* that the formulation of hypotheses required a creative process for which he had no theory at all.[25] But without the constraint of the real world to guide it, scientific imagination has no control—it can only fragment into thousands of possible accounts of reality, while the truth becomes all the more obscured. It is the severe confrontation of creative imagination with the constraint of observation and experiment that together produce the wonderful and dynamic field of the sciences.

No more does uncontrolled juxtaposition of paint, tones, words or forms produce good art, music, writing or architecture. Artistic imagination also requires the ordering constraint of *form* to create beauty that

[25] Karl Popper, *The Logic of Scientific Discovery*. London: Routledge, 2002.

can be transmitted and received. Whether sonnet or sonata, constraint plays as vital a role in art as does imagination. One of the famous paradoxes of all art is the essential role of form in liberating the imagination to create. Poet and thinker Paul Valéry remarked, 'A person is a poet if his imagination is stimulated by the difficulties inherent in his art and not if his imagination is dulled by them'. The difficulty of poetic constraint even correlates with the ultimate artistic potential of the poem; a truly great sonnet surpasses the greatest blank verse. Or as G.K. Chesterton playfully put it in the context of the visual arts, 'Art consists in limitation. The most beautiful part of every picture is the frame.' This works just as well in music. I will never forget the wonderful experience of singing in the (very lucky amateur) chorus for a performance of Vaughan Williams' *Sea Symphony* conducted by his great interpreter Vernon Handley. In a moment of respite during the rehearsal he started to explain why he loved conducting the excellent National Youth Orchestra of Great Britain. Part of his enthusiasm clearly stemmed from observing gifted young artists in the act of discovering what makes great art itself. 'I've got them well trained now', he boomed, with a twinkle in his eye. 'I call out, "What's the most important thing in life?"—and they all holler back, "ART". Then I ask, "And what's the most important thing in art?", and they shout, "FORM!".'

Recent psychological approaches to the role of form in generating creative art have identified the way that constraints generate the need to solve problems. Although this sounds more like a convergent than creatively divergent task, this is not the case when information is incomplete. In this case the mind starts to search more globally for solutions, taking more circuitous and less familiar pathways that initially seem to serve no purpose to attaining an aim, yet escaping from the narrow confines of a dead end. In a recent study[26] Patricia Stokes charts the long road that Braque and Picasso needed to travel to conceive the early twentieth-century art form of cubism. Their circuitous departure from representational art visited many intermediate experimental places before finding a satisfactory, yet surprising, home. We might be reminded of the counter-intuitive journey through the constrained chaos of the natural world that Job is taken on when he insists that what he needs is an answer to injustice, not a nature trail.

[26] Patricia D. Stokes, *Creativity from Constraint: The Psychology of Breakthrough*. New York: Springer, 2006.

Human minds seeking creatively to reconstruct the hidden processes below the surface of the natural world do so in a highly, but not over-constrained, way, reminiscent of the channelling double constraints of form and idea in art. Seen in this way, one could even claim for science a seat at the high table of the arts—for what demand on imagination could be greater than the commission to reimagine the world itself? What tighter form could be imposed upon that creativity than the very form of the world as it is? Braque and Picasso wanted to find a form of art that represented the world from more than one viewpoint simultaneously, that grasped the solidity and three-dimensionality of objects and landscapes even through the two-dimensional constraint of a canvas. Their journey strongly parallels that of science, which seems to set itself the impossible task of perceiving the internal structure of the world without leaving the confines of the human mind.

We will need more pliant disciplinary walls to create an academic environment that encourages students to explore and draw on consonant interdisciplinary ideas such as this. Thomas Bender[27] connects the desirable disciplinary aspects of vitality, objectivity and democracy with an inherently interdisciplinary character that he terms 'weakness' (without any pejorative connotations in his technical sense of the word). By this he means the openness to new movements and ideas that change the character of disciplines, but by evolving keeps them alive. We might re-express such 'disciplinary weakness' as a form of creativity with constraint. The creative energy inherent within a discipline works with the double constraints of its internal form and the threats and opportunities of the external world of ideas. He does not discuss the sciences, but his analysis applies both between them and more widely across our current faculty boundaries. The inherent 'weakness' of physics in his sense, for example, is clearly a strength that has enabled fresh views on such disparate and surprising fields as granular media (previously within the domain of engineering) and the molecular motors of muscle (previously biology). These movements are changing physics itself as well as reseeding those other disciplines with fresh questions and approaches. The 'strength' (in Bender's sense of inflexibility) of the engineering disciplines, by contrast, reinforced by the prescriptive demands of professional accrediting

[27] Thomas Bender, From academic knowledge to democratic knowledge. In *Values in Higher Education* (eds S. Robinson and C. Katulushi). Cardiff: Aureus Publishing, 2005.

bodies (at least in the UK), has impeded their development in such new directions. In looking for ways to realise 'creative weakness' in disciplinary boundaries, it hardly requires pointing out that even a minimal level of discourse between more widely separated disciplines requires the contemplative time that we have identified as one of our rapidly disappearing and precious resources.

The integration of knowledge and wisdom in the theological narrative we have followed gives hope that, by digging deep enough into the reasons why we explore our inner and outer worlds, we might realise a more connected enterprise of research. The heterogeneity of any theologically defined community points to this as well. If we can sustain the medical metaphor of 'healing', then it is as natural to recognise that the process requires the cooperation of multiple agents. So Rowan Williams urges that academic disciplines site themselves in proximity because, in spite of everything, they know they have 'something to do with one another'. In another theologically motivated lecture,[28] Nicholas Lash urges an academic grasp of the ultimately *connected* structure of the web of knowledge that underlies all disciplines: 'Notwithstanding the accelerated fragments of specialised academic activities, we trample in each other's territory, sing each other's songs, whether we want to or not'.

Surprisingly, we have discovered in our explorations a fresh way of moving on from the 'two cultures' paradigm that still ensnares us with its assumptions, not only in our universities but, more harmfully, throughout our education system. C.P. Snow's famous complaint,[29] and the vociferous mud-slinging (especially between Snow and the literary scholar F.R. Leavis) that ensued, has placed the centre of gravity of this debate firmly within a contest of knowledge claims, and their associated cultural value, of the arts and humanities on the one side and of the sciences on the other. Knowledge of the second law of thermodynamics is set up in rivalry for cultural value against familiarity with Shakespeare's sonnets. Or, in more recent guise of the 'science wars', physicists and social critical theorists have engaged in a series of disciplinary skirmishes whose weaponry includes bogus publications and books attacking the proponents of the other side as intellectual imposters.[30]

[28] N. Lash, *The Beginning and End of 'Religion'*. Cambridge: Cambridge University Press, 1996, pp. 112–31.

[29] C.P. Snow, *The Two Cultures*. Cambridge: Cambridge University Press, 1998.

[30] See, for example, Keith Parsons (ed.), *The Science Wars: Debating Scientific Knowledge and Technology*. Amherst, NY: Prometheus Books, 2003.

In contrast, the project of reconciling the human with the material by reimagining the world within the constraints of the mind, the theologically informed recognition that this is a sort of healing within reconciled communities, sets out a new framing of interdisciplinary relations. Rather than focusing on our differences and their relative valuation, we begin with our shared purpose, our shared experiences of creativity and constraints. These are exciting ideas to ground in masters' courses, and in due time the undergraduate student curriculum. Instead of contrasting our exclusive fields of expertise, we make progress by comparing and building on what is similar in our experiences and projects, illuminating the world in complementary ways. We may well find that the songs that the arts and the sciences both sing share some very ancient tunes and harmonies.

What We Do With the World: Narratives for Troubled Technologies

The twin tracks of knowledge and wisdom have refused to part company with each other throughout our journey. They have taught us to think of the relationship between science and technology in fresh ways, and then driven us to look more closely at the essential way that the separate disciplines we have created need each other. One way that a restored community of disciplines begins to act in fresh ways emerges from setting science once more into the frame of 'the love of wisdom to do with natural things'. For together, knowledge and wisdom now lead us to look within our new narrative resources for answers to a set of desperately urgent questions.

At no point in the history of *Homo sapiens* on planet Earth have we been less prepared to think through the consequences of our own actions for the world we inhabit and to understand the potential of our own abilities to change that world radically. The broad outlines of our predicament are well known. A human population growing from a sustainable population of a few hundred million towards figures beyond 10 billion is at the same time aspiring to a lifestyle which gobbles the planet's resources faster by far than the rate at which they can be replenished. Food and fresh water are rapidly becoming goods that humankind can no longer take for granted (in the case of food this has been so for over an eighth of the world's population for many years[31]). After hundreds of

[31] Food and Agriculture Organization of the United Nations, *Statistical Yearbook 2013*. Rome: UN, 2013.

millions of years sequestered in the Earth's crust, the thick early atmosphere, rich in carbon dioxide captured during the first aeons of plant life, is being returned through fossil-fuel burning to fuel this rampant economic growth, with evident but highly contested effects on the climate. The rate of species extinction has reached values exceeding by far that of any previous era, arising from both climate change and the radical reshaping of the landscape at our hands, including the removal of forest and ocean habitats. At the same time we fear the evolution of viral and bacterial strains resistant to our advanced pharmacology. Ironically, the very niches of survival for antibiotic-resistant bacteria that now threaten us have been created by those same medical advances. The natural world seems to respond to our technologies in deeply problematic ways that we did not anticipate.

At the same time we find ourselves strangely paralysed from taking effective action, or even exploring potential solutions with a confident care. This uncharacteristic impotence (in the face of the breath-taking potency of our technological development) takes different forms in different communities and within different questions. In Europe, the development of genetically modified organisms (GMOs) for consumption in the food chain has, since the early 1990s, been subject to the most severely regulated system of control in the world. This has allowed, in the last 20 years, the development of only one pest-resistant GM strain of any foodstuff within the European Union, contrasting markedly with the very liberal framework within the USA for this set of technologies, which has developed commercially for large global markets.

On the other hand, the development of stem-cell technologies with considerable medical application has, over the same period, enjoyed a permissive though carefully regulated licensing structure within the UK, while banned from public laboratories outright in the USA (during the Bush administration).

To take a third, less specific, example, the emergence of 'nanotechnologies' in the 1990s unleashed a confused public debate on both sides of the Atlantic on what new mechanisms of regulation should be devised to control possible harmful effects of manufactured particles only a few tens the size of atoms. Rather mundane applications, such as invisible suntan lotion, emerged in stark contrast to the terrible prophecies from senior public figures of a 'grey goo' spreading uncontrollably from laboratories, or to the overblown promises of 'smart nano-bots' injected into cancer patients on search-and-kill missions against tumours.

Finally, the global charade of climate-change conferences since the Kyoto accord testifies to the bluster that passes for concerted political action. Carbon emission reductions are talked about, but not implemented, at the behest of short-term economic gain by the nations which imagine that they have most to lose by radical action to transfer energy generation to low-carbon economies. Equally stalled are the 'sticking plaster' technologies of geo-engineering which seek to reverse global warming by means other than reducing carbon emissions, such as the introduction into the upper atmosphere of reflective aerosols.

These examples are not meant to be evaluative judgements of right or wrong—rather they point to a lack of consistent ability to make public and political decisions around a class of powerful technical developments. We might call them 'troubled technologies', for they seem to share a number of challenges in common: (i) they may contain great potential advantage (e.g. pest-resistant crops, regenerated organs for transplants, global mass transportation); (ii) they bring unknown or partially known risks before a significant evaluation has occurred (e.g. the resilience of GMOs to other infections or the risks associated with cross-fertilisation, unknown side-effects of artificial organs, climate change itself); (iii) they tend to generate, sometimes in a subset of communities only, an atmosphere of fear transmitted and developed in a media suspicious of science (e.g. the 'Frankenfoods' translated metaphor in the case of GMOs and 'playing God' in the case of artificial organs or tissue generated from stem cells); and (iv) the troubled technologies seem to touch a nerve of disquiet at a deep human level—they carry the aura of trespass, of the crossing of the threshold (it is in part this aspect that fuels the public narratives of fear).

Careful research into the public debates around these technologies has yielded some fascinating discoveries—there are strong indications that the cross-talk around troubled technologies is not all that it seems. The specific example of nanotechnology is an instructive starting point. A major 3-year European research project between the universities of Durham (UK), Darmstadt (Germany), Twente (The Netherlands) and Coimbra (Portugal) from 2006 to 2009 explored in detail the discussion on nanotechnology at the levels it had achieved in Europe at that time. Ostensibly, the chief political question was that of unknown toxicological risk around nanoparticles in the environment. Proponents of nanotechnology development tended to argue that the new materials posed no qualitatively new risks that could not be controlled by existing frameworks. Opponents, sometimes referring to past cases of unknown risk, such as asbestos fibres, remained unconvinced and tended to emphasise the qualitatively new technical features of nanoparticles such as their similarity in scale to viruses, or the open-ended and uncontrolled character of a technology that made claims to be 'bio-mimetic'. When this includes the self-assembly and adaptivity of molecular structures to explore configurations beyond any previously known, then new landscapes of unknown risk present themselves.

As well as extensive documentary analysis, the large team of social scientists and technologists used conversational and discursive techniques with people from many different backgrounds to peer beneath the superficial substance of regulatory debate. Their project report, *Recovering Responsibility*,[32] tells a very different story from that of the claims and counter-claims within official reports of public consultations. Tellingly for our theme, it describes the unearthing of underlying narratives of suspicion—stories and themes that influence and permeate the debate, without necessarily surfacing. As identified by philosopher Jean-Pierre Dupuy,[33] they draw on both ancient and modern myths, and create an undertow to discussion of troubled technologies that, if unrecognised, renders effective public consultation impossible. The research team labelled the narratives:

1. Be careful what you wish for—the narrative of Desire
2. Pandora's Box—the narrative of Evil and Hope
3. Messing with Nature—the narrative of the Sacred
4. Kept in the Dark—the narrative of Alienation
5. The rich get richer and the poor get poorer—the narrative of Exploitation.

The first three Dupuy unites in an 'ancient meta-story', the last two in a 'modern meta-story'. It is at first rather astonishing to find as superficially modern a set of ideas as nanotechnology awakening such a powerful set of ancient stories, but in the light of our analysis that the problematic engagement of the human with the material is actually very ancient, and embedded in the discourse of sacred texts and the stories of their communities, it becomes less so.

[32] Sarah Davies, Phil Macnaghten and Matthew Kearnes (eds), *Reconfiguring Responsibility: Deepening Debate on Nanotechnology*. Durham: Durham University, 2009.

[33] J.-P. Dupuy, The narratology of lay ethics, *Nanoethics* 2010, **4**, 153–70.

New technologies, especially those whose functions are hidden away at the invisible molecular scale, promise much, and have made exaggerated claims of benefits: longer, healthier lives at low cost, self-repairing materials and machines, built-in sources of energy. But such hubris elicits memories of overpromising: behavioural engineering, nuclear power, cures for cancer. The fear of unforeseen consequences drives the conclusion that 'desire' makes blind: a vision of the future that entrances advocates of a new technology by the same measure makes them unwilling, or unable, to think through the possible risks. Claims that new technologies can be controlled meet with concerns that humans have always been confronted with limits to their control of nature's processes.

The story of Pandora's box enters at this point, for this tale of the seductive power of the hidden speaks across the ages to our power to unlock the twinned histories of trouble and hope. As a foil to the narrative of 'desire' is the narrative of the fear of harms, and in particular of the irreversible kind. The swarm of trouble bursting from the box might well remind us of an expanding cloud of gas molecules following the second law of thermodynamics into irreversible disorder. There is one sense in which irreversibility always accompanies technology—communally held knowledge is very hard to forget once it has been learned. Now that the knowledge to build a thermonuclear explosive device rests in a number of places and within many human minds, it is hard to envisage how, barring global catastrophe, that knowledge would ever be lost.

The nanotechnological counterpart of the Durham study identifies irreversibility both in knowledge gained and in the 'release into the environment' of nanoparticles. Pandora also released hope from her casket—in the original myth usually read as a positive and counteracting good. However, as Dupuy points out, hope can be dangerous: it can drive a course of action onwards beyond the point at which a dispassionate risk analysis would have recommended a halt. Political language such as 'responsible development of nanotechnology' disguises an underlying hope that is seductive and potentially irrational.

The third 'ancient narrative' is a fascinating and perplexing one. Why would a secular age develop a storyline that warns us away from 'messing with nature' because of its sacred qualities? The secularisation of thought and society has been charted, in the last century, in social theory from Emile Durkheim and in political philosophy from Hannah

Arendt.[34] The narrative is a familiar one: in the face of increased global connectivity religion needs to come to terms with pluralism; in the face of science with secular explanation of nature; in the face of politics with power drawn from the secular not the sacred. The power of all three progressively marginalises religion into irrelevance and ultimately disappearance. Even the more recent social analyses driven by the manifest persistence of religious thought into the modern world, such as that of Jürgen Habermas,[35] have approached religions as minority communities that need to come to terms with the secularisation of the majority using their own resources, or face even further marginalisation.

The ancient narrative of trespassing on the sacred, or of 'playing God', is not, however, confined to any explicit religious community or tradition. It is very widespread, though it seems to surface only when called upon by public debates of troubled technologies. Peter Berger, in his account of 'desecularisation',[36] identifies only two exceptions to the persistence of religious influence, two communities where the traditional twentieth-century secularisation theories still play out: those of geographical Western Europe and demographical academic communities. But even here the nerve is still raw according to the nanotechnology study. A fascinating example of how the 'messing with nature' narrative is articulated, in the context of another troubled technology, is provided by the process of fracture recovery of coal gas from near-surface shales known as 'fracking':

> *In ancient times, people believed that inclement weather came directly from a divine source: Whether it be Gods, Goddesses, or just the 'spirit of the planet', we have always arranged sacrificial offerings and desperately tried to appease whichever deity has punished us for our wickedness. Although we have somewhat 'grown out of' this concept of divine retribution for sin, we kind of have to admit that we have become sinful in our collective attempts to thwart nature and impose our will upon it.*[37]

[34] Hannah Arendt, *The Human Condition*. Chicago: University of Chicago Press, 1958, p. 314.

[35] Jürgen Habermas, Religion in the public sphere, *European Journal of Philosophy* 2006, **14**, 1–25.

[36] Peter L. Berger (ed.), *The Desecularization of the World: Resurgent Religion and World Politics*, Washington, DC: Ethics and Public Policy Centre and Eerdmans, 1999.

[37] Chris and Sheree Geo, *Mother Nature Becoming More Irritated As Even Geoengineering Scientists Admit They Might Be Making Things Worse... But They'll Keep Doing It, Anyway*. <http://www.geoengineeringwatch.org/> (accessed 13 March 2013).

Sacred spaces, it seems, are not confined to the human genome, but extend to the subsurface of the planet we live on. We 'mess with' these at our peril, so the narrative goes. These debates draw energy from a contested view of nature: between a given order that humans tamper with at their peril and a field of opportunity imperfectly suited to our needs, but pliable in the face of technology. When the inefficiencies of blind evolution are set against the real possibility of human design, the stage for hubris is set, and with it the dangers of assumptions that our capacity for control is boundless.

The challenge is to create a proper and functioning balance between the technological and cautionary tendencies in public debate. It is in the realm of the ancient narrative of the sacred that we currently suffer the greatest dissonance and unmatched debate—the self-confession of 'growing out of' old stories points to the crevasse into which current conversations all too frequently disappear, sending opposed communities into the safe zones of their own websites. Nature mythology poorly articulated, and technical geo-engineering simply make very poor conversational partners.

The fourth narrative of being 'kept in the dark' is at first sight, as Dupuy observes, a more modern one, speaking of asymmetries in political power between the governing and the governed. We know now, however, that it also draws on very old stories of ignorance and knowledge as well. There is a parallel structure of darkness in the natural and human spheres that makes this narrative so problematical. Nature is hidden in the darkness of ignorance—this is part of the painful relationship with humankind. It is *mirabile dictu*, possible for us to illuminate the physical world's dark spaces with observation and mind, but this can as easily generate a new 'priesthood' and new boundaries of ignorance as it can break them down, even when this happens inadvertently. If people are being 'kept in the dark', however, this is a deliberate act of power creating powerlessness through ignorance.

The fifth narrative of 'the rich get richer and the poor get poorer' extends the fourth into its consequences for creation and distribution of wealth. With exclusion comes lack of access to the benefits of knowledge, and, worse, unequal exposure to their harmful consequences. Elevated publicly stated goals of an equable world, such as enshrined in the United Nations' Millennium Goals,[38] are insufficiently linked to

[38] United Nations General Assembly, Fifty-fifth session agenda item 60(b), A/RES/55/2. UN, 2000. <http://www.un.org/millennium/declaration/ares552e.pdf>.

strategic programmes of delivery to make them credible. The ethical and strategic vacuum gives place to this story of injustice, and a suspicion that technology is indifferent to the realities of human suffering. It is fed by the knowledge that commercial interests are always inextricably bound up in projects ostensibly aimed at the public good. This fifth narrative has, for example, been especially prevalent in the resistance to GM crops in India, where scientific discussions of possible health risks are routinely undermined by the identification of 'conflicts of interest' arising from funding. This narrative is, perhaps, not so very modern after all; nature, suffering and injustice are themes within a very old song with which we have now become familiar.

The European nanotechnology study is interesting, not only because it begins to make progress in perceiving why our newest technologies are so troubled but also through its unearthing of the fundamental importance of underlying narrative. It illustrates in a sharply practical way the findings of this book, but by a completely different route. By close-reading theological sources that speak of our relationship with the natural world, and through the current experience of doing science today, we have outlined a narrative theology of science that promises to resolve many of its painful nerve endings felt in developed cultures today. We have found resources that help re-situate science within the long story of human culture, and in particular challenge superficial contemporary assumptions about what the relation between science and religion might be. Now, in the social-scientific analysis of troubled (nano-) technologies we see the problem from the other end: here there are (at least) five ancient narratives coiling around a resistance to new science and new technology. They highlight in the most lurid possible contrast that science itself has no such source to draw on—*there is a narrative vacuum where the story of science in human relationship with nature needs to be told.* What might happen to public debate on contentious science and technology if there were an active ancient narrative that was more neutral, or even positive, in its recounting of 'love of wisdom to do with natural things'? What would a public debate on troubled technologies look like were all these ancient narratives to be made explicit, and to have their resources engaged in figuring our future path?

Let us trace the beginnings of a discussion between the five narratives of the *Reconfiguring Responsibility* report, and our human, historical, relational and participative theology of science. 'Be careful what you wish for', while carrying a health warning against boundless desire, is also shot-through with resignation and conservativism. It fails, ultimately, to recognise the unacceptably painful current state of our ignorance and fear in the face of the non-human material world. Furthermore, our situation of scientific research *within* rather than *against* a theological context does not immediately supply any predetermined ethical boundaries for the subject of investigation or interrogation of nature. It recognises, even celebrates, the extraordinary ability of the human mind to sift the causes of phenomena, and also to alter them. The final biblical vision of the 'new creation' of the Revelation to John is, after all, a *city* rather than some Arcadian vision of rustic simplicity. While guarding against a ruthless and exploitative domination of nature, the command at Eden, the agricultural engagement of Isaiah, the beckoning invitation to wisdom of the Lord's answer to Job are all encouragements to wish for very great things indeed. This is perhaps the greatest surprise of a relational and participative theology of science—it is not at all conservative in its estimation of what humanity is capable of, nor in what it ought to aim at. There is a broken relationship with nature that needs the deepest of commitments—the language of covenant has not, as we have seen, been too strong.

The double contents of Pandora's box of trouble and hope are consequentially seen in a new light. The message of Job is that chaos is part of the fruitfulness of creation; we cannot hope to control it any more than we can bridle Leviathan, but by understanding we might channel it. Indeed new structures can arise when we do—the 'beginning of wisdom' is not to double-lock the casket of our ignorance, but to seek the 'fear of the Lord', where this is understood to be a participation in a creator's deep insight into the structure of what he has made. As for hope, the jagged edge of hoping for too much without a proper tension with risk is tempered by situating our science and technology within a story of participative healing. By no means a simple mandate for a thoughtless pursuit of technical fixes, the theological narrative recognises that there is a past and a future to our relationship with nature—and that there is a place for both warning and hope as companions. From a Christian standpoint the resolutely physical embedding of the very idea of hope itself, the Easter story, cannot go unremarked. Not only is the resurrection a tangibly substantial, materially embedded sign of hope, it is a future-pointing one. An over-'spiritualisation' of Christian theology for centuries has deflected attention time and again from this greatest possible sign that physical embodiment matters, and that hope for a reconciliation and healing of humanity with God is bound up with the reconciliation and healing of physical nature as well. St Paul and St John understood it, and expressed it in the powerful metaphors of a world groaning in childbirth and a healed world that most resembles a city and its garden.

Within this story, 'messing with nature' is more of a mandate than a forbidden and dangerous activity. We have seen how bringing a participative 'ministry of reconciliation' alongside the story of a broken relationship with creation brings into focus the idea of a servant–priesthood of mediators between the human and non-human world. Within a religious tradition, science becomes a holy task (another immense surprise), but even from a secular anthropological perspective the theological work helps to resolve an artificial frontier between human and material. This is the chasm of ignorance and fear that leads to the opposite harms of complacent yearning for an imaginary and perfect past natural state on the one hand, and a short-sighted and human-centred exploitation of nature on the other. A 'holy messing' with nature retains the playful but perceptive picture of Wisdom, the little girl in Proverbs collaborating with the creator in forming the Earth, and meets the challenge to Job to answer the great questions of the cosmos. It recognises that we are ourselves made of the dust of the Earth, but endowed with the responsibility of creativity. Returning to the language of boundaries and journeys, a participative theology of science may create no fenced-off areas with 'thou shalt not trespass' signs, but it does keep a compass in its hand along with a clear direction of travel.

There is no place in our participative narrative of wisdom for the power-play of 'keeping in the dark'. On the contrary, the purpose of reconciling with nature is a universal one. Individual tasks and abilities naturally differ, but an overstratification of science prevents a damaging alienation from physicality at the personal level, and a dangerously disconnected debate in the public and political arena. The consequences of a widespread failure to understand the process and context of research are widely discussed, particularly in terms of the public perception of risk. Whether the issue is nuclear waste disposal, GM foods or animal cloning, we are beginning to talk of the validity of ethical viewpoints of scientists and 'non-scientists'.[39] Recognising valid and disparate grounds of choice in the challenging grand issues before us is an essential step, but, in attempting to solve global issues rapidly, discussion restricted to this level sidesteps the *local* spadework of generating a shared recognition

[39] D. Burke, Assessing risk: science or art?, *Science and Christian Belief* 2004, **16**, 27–44.

that science is at the core of human creativity. We need perhaps to return to Macrina's bedside to learn this afresh: there is a direct link between her playful, childlike and perceptive reasoning on the property of air to an appreciation of how an organism's genes change naturally, and may be changed artificially. Both build on simple images accessible to anyone, while calling on a challenging degree of contemplation (one that, after all, pointed Gregory of Nyssa to identify the reality of an independent mind). The language we use will be crucial.

The narrative of poverty and justice strongly engages with another consequence of our theology of science—the reappraisal of its value. It is not surprising that economic rationales for research are the only ones on the table when no others are offered, so become fuel for suspicion in debates around troubled technologies. The nexus of science—the material world and the cry for justice—is, of course, right at the heart of our most central textual source of the book of Job. If ever readers were at a loss to understand the connection between the injustices of Job's complaints and the theme of nature hidden and revealed, then the criticism that new technologies simply fuel 'the rich getting richer and the poor poorer' shows that Job's cries continue to this day. Losing sight of what science is for is the first step to stripping it of its values, and the disabling of a vital tool in the formulation of science ethics. The themes of reconciliation, communities of shared values and a primary engagement with the world are very ancient. But at the same time they speak urgently to our present predicament of public unease with science, and in particular with its unbalanced connection to economic values and minority interests.

To realise the vision of an ethical research process, democratically shared, living and vulnerable, in the face of the severe challenges we have also identified, requires a faith in the scientific community and wider academy of research that is increasingly hard to find. Yet the relational and reconciliatory task at its core that we have identified needs just the form of faith proposed by Rowan Williams[40] in his own discussion of faith in the university: '. . . a commitment to the belief that our life is more than a struggle between a creative ego, individual or collective, and a lot of raw material; it trusts that there is a possible reconciliation ("atonement") between human selves and their world'.

[40] R. Williams, Faith in the university. In *Values in Higher Education* (eds S. Robinson and C. Katulushi). Cardiff: Aureus Publishing, 2005.

A 'love of wisdom to do with natural things' as a deeply human story with a long history, which is also embedded in many religious traditions, and resolutely in the Judaeo-Christian one, promises some exciting practical consequences if entertained within and beneath public discourse. At the very least it will counterbalance, challenge and inform other ancient myths that are determining the direction of current debate. More than that, it may inspire and support new communities of engagement and a strategy for meeting laudable goals of global justice with participative technologies. It might just move us beyond suspicion, through a just application of our privileged knowledge, to an engagement with the planet's resources marked more by wisdom than folly.

What We Do With the World: the Politics of Nature and a New Environmentalism

The 'troubled technologies' we have just explored show how a well-founded and human story around the purpose of science might help navigate some important yet thorny technological questions and improve the health and effectiveness of public ethical debate. But working through them has felt a little like stumbling through a darkened cellar, flashlight in hand, shining it on each box as we unpack one at a time. Their contents suggest that there is a wider panorama to see if we can only find the light switch.

The bigger picture is all around us—the very impression that we are still stumbling in the dark is itself an indicator of our wider problem. There is no better example of the phenomenon of 'marginalisation' than the current public position of what we call the 'environment' (the very word reflects the marginal). Current affairs webpages place 'science and environment' towards the bottom of their tab-list; it creeps in as an element of public justification in some space-science outreach programmes. The occasional documentary or newspaper position piece is timed to appear around the annual climate change summit, but typically constitutes a few pages of predictable hand-wringing. 'Green' political parties have come and (more or less) gone. In spite of growing consensus around anthropogenic climate change, shortages of food and water for a growing global population, the evolution of resistant strains of bacteria, accelerating species extinction and the loss of biodiversity, none of this captures political energy or will at all comparable to the topics of rogue states, international terrorism, immigration control and presidential elections.

We can continue the list of 'media-friendly' topics at ease. There is no shortage of public political and electoral drive around issues of compromised political relationships and no shortage of headline spaces for contested opinion and debate in old and new media alike, but that bandwidth is simply not commanded by the equally pressing flawed relationship of humankind with nature. Our failure to get any grip on the issue of our own damaged environment is embarrassing to us; the way we marginalise it is reminiscent of an old family feud which people prefer to ignore than face up to. It is, increasingly literally, the elephant (a rapidly decreasing number of elephants) in the room of our political conversation. The fresh urgency of the environmental movement of the 1960s and 1970s has been somehow dissipated, diverted into dismissible extreme views or steamrollered flat by national and corporate economic interests. Worse—it has become publically *dull*. Why has political energy been frozen out of the environment?

One leading contemporary commentator whose interest in the 'politics of nature'[41] has not been marginalised is the French thinker Bruno Latour. In a recent edited volume[42] he explores just this question—with conclusions that are remarkably resonant with our own. They break down into four findings, in his own words: 'a stifling belief in the existence of Nature to be protected; a particular conception of Science; a limited gamut of emotions in politics; and finally the direction these give to the arrow of time'. This is a grand, overarching critique of the politics of nature, but, even so, it homes onto the same narrative analysis as did the specific nanotechnology study we examined in the last section. His identification of the 'stifling' move to withdraw all human corruption from a 'nature' that should be maintained in some pristine condition is none other than the 'messing with sacred nature' narrative by another name. Latour extracts the self-contradictory structure of this story of the 'golden age'—for all 'protection' is by human construction in any case, even if there were any such natural domains left to protect. Nature reserves are artificial by definition (and wildlife commonly disrespectful of their

[41] Bruno Latour, *Politics of Nature: How to Bring the Sciences into Democracy*. Transl. Catherine Porter. Cambridge, MA: Harvard University Press, 2004.

[42] Bruno Latour, 'It's development, stupid!' or: how to modernize modernization. In *Postenvironmentalism* (ed. J. Proctor). Cambridge, MA: MIT Press, 2008.

boundaries). We are already so intimately connected with our natural environment that withdrawal from it, a return to a non-technological civilisation, is no longer an option. We are—and need to be—committed.

But the alternative 'modernist' trajectory is no less problematic. There the story is an overcoming of nature with control. We disengage from our environment, not through an 'environmentalist' dream of withdrawal from the sanctuary but through a technological domination. Here Latour revisits the narrative of Pandora's box, because such a modernist hope is dashed on the rocks of the same increasingly deep and problematic entangling with the world that prevents withdrawal. Nature does not respond mildly to an attempt to control or dominate. Neither narrative works—both start with fundamentally misguided notions of the geometries and constraints of our relationship with nature. The old story of the painful divide between the human and the material—painful because the two can never really be divided at all—returns to stifle us into inaction. We cannot go back, yet neither of the two routes presented as the only alternatives solves the family feud with our world.

Latour's critique of the conception of science is equally resonant with the flawed view of a 'scientific priesthood' we have already explored. Political action on scientific decisions is as paralysed by disagreement as it is by disengagement. Not every expert agrees that blood transfusion might transmit the AIDS virus—so we wait in inaction that condemns children to infection. There is no uniform view on the future trajectory of global warming and its connection with human release of carbon dioxide—so we meet and talk, but do not implement. This is the 'kept in the dark' narrative with a twist—the political and public community self-imposes ignorance by demanding that scientists behave as a conclave, reading the same script and praying the same prayers, until the white smoke of expert agreement is released. The political life-blood of a communally possessed and confident debate, widely shared and energised, respecting where specialist knowledge lies, but challenged within a participating lay public, is simply not yet flowing in our national and international veins.

At the close of his contribution to *Postenvironmentalism*, Latour makes an extraordinary move—one that meets our own journey head on. He calls for a re-examination of the connection between mastery, technology and *theology* as a route out of the environmental impasse. We have not yet remarked that the ancient narratives unearthed by the nanotechnology project, and reflected in Latour's, are all implicitly or explicitly pagan, though we have seen how they might be met with, and transformed by, the more positive themes of a Judaeo-Christian 'ancient narrative' of nature. So when he refers to the Christological theme of the creator who takes the responsibility to engage with even an errant creation to the point of crucifixion, the contrast with the disempowering and risk-averse narratives of 'being careful what you wish for', Pandora, sacred nature and the rest—could not be starker.

The theological wisdom tradition we have been following, especially in the way that it entangles with the story of science itself, has brought us to the same point that Latour reaches from the standpoint of political philosophy. One identifies the need, the other the motivation and resource, for a re-engagement with the material world, and an acknowledgement that one unavoidable consequence of being human is that we have, in the terms of the book of Job, a 'covenant with the rocks'. This extraordinarily powerful collision of metaphors surely points to the balanced and responsible sense of 'mastery' that Latour urges that we differentiate from the overtones of exploitative dominance. More is true—if we take one by one the strands of the 'theology of science' that we teased out in Chapter 7 of our biblical nature trail, it begins to look as though they might be woven into the story, becoming the missing narrative that Latour wants to hear told.

The thread of *linear history* tells us that we are not at equilibrium, we are on a journey from ignorance to knowledge (whether we will it or no) and wisdom (if we so choose), a history of questioning nature and engaging with it. It provides a frame of long timescales to the current bewildering speed of science and technology, reminding us that, while a steering wheel and a foot brake are controls we might exercise, turning off the engine of science and coming to a standstill would amount to denying our own humanity. The recognition of *human aptitude* supplies the necessary energy of a qualified optimism that we do, after all, have the mental and social capacities to manage our relationship with nature away from harms and into fruitfulness. The endowment of human abilities that attain to the 'co-creational' might sound protean or over-reaching in a nihilistic universe, but it is both appropriate and necessary in a cosmos shot through with meaning. For the thread we called *deep wisdom* turns human aptitude away from the modernist narrative of dominance towards a balanced and humble 'mastery' in the truly participative sense.

Wisdom recognises that nature needs the violent energies of both *order and chaos* to give birth to life in the first place. Development of natural form and structure likewise rides upon the waves of random exploration with the space of possibilities. So wisdom urges us towards a technology of direction and management, a softer kind of mastery.[43] It steers us away from technologies that confront natural forces by walling them in or reversing them. Think of Job's desert floods finding their way through channels rather than breaking through hopeless attempts to dam or contain them. But think also of the miner digging for minerals concentrated into seams by processes of volcanism or crustal folding, or Isaiah's experienced planter and tender of crops working alongside nature's tendency to select for reproductive capacity. *The ambiguity of problems and pain* warns us that taking this road of engagement with nature, which bridges between our potency for fruitfulness and our potential to damage, will be difficult. This type of mastery—one could almost call it paradoxically 'servant mastery'—calls on one who engages to suffer. The covenantal imagery in Job, no less than the weight of painful ignorance brought onto Job's shoulders by the Lord's answer, points to the expectation that this relationship, like all relationships, either freezes or makes progress through pain as well as joy.

Looking at our technological impasses in this light can release some of the stalemates we experience. We do our best to anticipate side-effects from new drugs, or, for that matter, from technologies for sustainable energy—but the certainty that there *will* be harmful mistakes is not a reason for inaction. It is an incentive to consult widely, to experiment gradually and openly, for transparent public governance of new science—in short for application of practical wisdom. It tells us to continue asking *questions*, the fundamental units of interaction between the human and the material. The great meta-question—'Where can wisdom be found'—is granted, in the Lord's answer to Job, the echo of a myriad detailed, probing questions into nature that together take the book's readers from ignorance through the world we perceive yet do not understand, beyond into realms of creation we do not even know of. Living with questions is characteristic of wisdom; insisting on answers to everything now, before we take a step forward, is not. But such a way of life tells us to go on

[43] The relevance of such a theologically guided wisdom for the future regulation of biotechnology has been argued forcibly by Celia Deane-Drummond in *Creation through Wisdom: Theology and the New Biology*. Edinburgh: T&T Clarke, 2000.

working with nature in sickness and in health—because at its root this relationship is one of *love*. To love the world, to be engaged to it, covenanted to it, entangled with it, to grow up together with it, *even though we understand it now only dimly*, is at the heart of what it means to be human.

We have summarised this theology of science as a story of *reconciliation by participation*. It begins to look as though this might be the missing narrative that takes environmentalism on, not just theoretically or theologically but practically. If our exploration of ancient wisdom and its hidden nourishment of the story we now call science is not misguided, then its subterranean stream is one that we need to tap. If the narratives of 'messing with sacred nature' on the one hand and technological utopia on the other can propagate in such a way that they become controlling (and competing) narratives in real public debate and policy decisions, then so can this one. New narratives (and renewed old ones) have a reproduction rate, thanks to new media, vastly greater than even a single generation ago—it is not a forlorn hope that the storylines of science might really recover a deeply human foundation and begin to feed debate in a positive way. If environmentalism needs renewed energy, then the source that powers creation itself is a good place to dig.

Personal Stories and Science Therapy?

One of the most moving encounters I have experienced while talking about science in different public settings came about in the West Yorkshire town of Dewsbury. I cannot recall now how it was that I was presenting an evening on the science of polymers to a Women's Institute meeting there. I had all the 'props'—buckets of slimy liquids to demonstrate with, computer graphics of entangled molecular strings in motion, and stories of how coming to grips with the molecular structure itself was now helping industry develop new materials (the region has a strong history in textile manufacture). I do remember one elderly lady who, in spite of my fumblings and hesitations around what would really interest this group, held a look of rapt interest throughout the talk. She never once seemed to lose concentration in nearly an hour, and come question time I was not surprised to find who was keenest to find out more. The wonderful, insightful questions came pouring out, and after a while others joined in and we had a rather serious seminar.

We learned her personal story too. Aged 15 she had left school and gone straight to work in the textile mills. Everything around her fascinated her, and she said that she always wanted to know more: why did the stretched yarn vibrate in such a way when the shuttle came past? Why did the shade of colour of the dye on the fibre differ from its appearance in the liquid? But her questions were never answered, 'Oh Betty, stop your questions and get on with your work!' was the response she usually received. For many people the relentless put-downs might have snuffed out her curiosity—but she had kept it alive for 50 years. The talk and discussion that evening acted as a long-awaited affirmation that there was nothing wrong with her interest or her questions. She actually shed a tear or two—and she was not the only one.

We might hope that, were Betty at school today, there would be nothing to prevent her curiosity leading to study of science at the highest level, if that is what she wanted. But in spite of wider access to education, we know that we still manage to keep this natural delight in questioning the physical world from so many, convincing them that it is only for the gifted (and freakish) few. We began by deconstructing the word 'science' with its Latin cognates of knowledge claims, and overpainted by its relatively recent use in the word 'scientist' (unknown before 1832) in the dominant colours of Victorian industrialisation. I recall that the occasion of the Dewsbury Women's Institute evening was one of the first in which I suggested that we all experiment by thinking in terms of the older, Greek cognate of 'natural philosophy' or 'love of wisdom of natural things'. People there began to smile (especially Betty) when thinking in this way about a realm that has previously generated fear or incomprehension. We have come across other personal stories of science—I recall my neighbour's sudden realisation that he understood for the first time how it is that the Moon shows different phases during the month, and his consequent ability to imagine for the first time the three-dimensional relations between Sun, Moon and Earth even as he looked up, with fresh eyes, into the sky.

Such small steps as this can begin a personal reconciliation with common aspects of the physical world: a realisation of why a rainbow appears when it does, the reasons food smells nice when it is cooked. These examples, as much as Macrina's comforting her grieving brother by rehearsing the cause of the Moon's phases 16 centuries ago, point to a personal consequence of a long human story of science. If communities and nations are suffering from a broken relationship with the material world, then might this not also be true of individuals, especially in light of the political and social framings of science that clash so starkly with the theological path we have followed? If an author like Bill Bryson[44] can experience a sudden and overwhelming self-doubt in the face of an unreconciled ignorance of how nature works, surely the buried need that surfaced then might be troubling many others, but go yet unrecognised?

The very strangeness that the notion of 'science therapy' elicits (it would have sounded ridiculous in Chapter 1) indicates that there is thinking space to explore here. 'Surely only art can go some way towards reconciling us . . .' said George Steiner—and indeed it is true that we have found 'art therapy', 'music therapy', 'drama therapy' and the like both rehabilitating and enriching. If a reintroduction to the activity of representing both inner and outer worlds in paint, music and drama can help to heal minds, what hope might there be for a participation in a gentle and contemplative science in restoring a broken or misunderstood relationship with the physical world?

A theology of science acts to reconcile in any case, as we have seen, the differentiated disciplines that we distinguish in our schools and universities (and recall again that in almost any European language apart from English, there is no word that translates 'science' with as narrow a meaning as it carries in the UK or USA). Science therapy would bear no resemblance to the experience of a classroom or teaching laboratory—more fruitful would be a design that draws on the playfulness of wisdom's exploration of the material (of Proverbs) and the open-ended generation of questions (of Job). A structured 'sandpit for grown-ups' might better describe the approach. Internet-based access to information releases enormous potential as a part of the personal exploration, but does not substitute for material manipulation of lights, objects, fluids. 'Looking beneath the surface' plays an essential part—a microscope needs to be at hand, not necessarily to supply answers, but primarily to do for participants what the Lord's answer does for Job when delineating in powerful detail the design of Leviathan. A strangely infrequent comment on Job's story is that, in spite of the claimed inadequacy of his whirlwind tour in those marvellous chapters, in any case as an answer to his complaint, the effect is one of reconciliation and reconstitution. Saying that 'science heals' might mean very much more than the discovery

[44] See Chapter 1.

and manufacture of medicines. Radical, even absurd, as it may sound, the music and poetry of looking deeply into nature and repatterning our minds to embrace what we see in the outer world might itself be the therapy suited to some of our inner troubles as well.

We are led to a last 'Joban' question: by contemplating, sharing and nurturing a deeper perception into the still-unknown fields of chaos and order of our universe, can we learn to heal the troubled relationship with our world? Can we learn what 'loving wisdom of nature' might mean? Do we have the wisdom to count the clouds?

Epilogue: a Parable for Science

I have always been deeply impressed by the extraordinary encounter of Jesus with a Roman Centurion, recorded in Luke's gospel (chapter 7):

> *When Jesus had finished saying all this to the people who were listening, he entered Capernaum. There a centurion's servant, whom his master valued highly, was sick and about to die. The centurion heard of Jesus and sent some elders of the Jews to him, asking him to come and heal his servant. When they came to Jesus, they pleaded earnestly with him, 'This man deserves to have you do this, because he loves our nation and has built our synagogue.' So Jesus went with them.*
>
> *He was not far from the house when the centurion sent friends to say to him: 'Lord, don't trouble yourself, for I do not deserve to have you come under my roof. That is why I did not even consider myself worthy to come to you. But say the word, and my servant will be healed. For I myself am a man under authority, with soldiers under me. I tell this one, "Go," and he goes; and that one, "Come," and he comes. I say to my servant, "Do this," and he does it.'*
>
> *When Jesus heard this, he was amazed at him, and turning to the crowd following him, he said, 'I tell you, I have not found such great faith even in Israel.' Then the men who had been sent returned to the house and found the servant well.*

I love this story. It shocks and surprises at every turn—a Roman Centurion builds a synagogue and applies to a Jewish rabbi on behalf of his servant. He commands a fond respect among the local people, although this is in an occupied country in which he represents the oppressor. He makes a supplementary request that Jesus not actually attend—and we are confronted by Jesus' own amazement (this is the only occasion ascribing that emotion to him in the gospels). But it is the *reason* that Jesus declares the centurion to have faith that is really striking. The soldier's belief that Jesus can heal his servant is not the surprise—for that much is clear right at the beginning. No, what Jesus calls 'great faith' is the *understanding of true authority* that the centurion demonstrates. He knows that his authority comes not from himself but from his superior in command—crucially he does not extract from it a licence to dominate, but the authority to care, to make fruitful, to bring about reconciliation. His servant's healing serves as a sort of reflection of the healing he works for all around him. Jesus is amazed—and calls this kind, strong, creative, healing wisdom 'faith'.

The nameless centurion might serve as an icon for us, who have tried to understand where the authority we call 'science' might lie in the human story. Whether we align ourselves with this—Christian—tradition or not, it is clear that the ability to do science, to deploy the 'love of wisdom to do with natural things', endows us with extraordinary authority and responsibility. The centurion was able to use his authority in service rather than domination to create reconciliation rather than antagonism, to invoke power to heal rather than to hurt. His humility enabled him to engage the community around him in the project, to celebrate together, to share in suffering and in achievement. If our reading of the long story of science is right, then this is where we also stand, and similar challenges lie before us. Just as there are two very different ways of being a military officer in an occupied country, there are two very different ways of being a community that does science. Can we choose the way, in wisdom, that deserves to be called 'great faith'?

Bibliography

Aggeli, M., M. Bell, N. Boden, J.N. Keen, P.F. Knowles, T.C.B. McLeish, M. Pitkeathly and S.E. Radford, Responsive gels formed by the spontaneous self-assembly of peptides into polymeric β-sheet tapes, *Nature* 1997, **386**, 259–62.

Anderson, P., More is different, *Science* 1972, **177**, 393–6.

Appleyard, B., *Understanding the Present: Science and the Soul of Modern Man*. New York: Doubleday, 1992.

Arendt, H., *The Human Condition*. Chicago: University of Chicago Press, 1958, p. 314.

Atkins, P., Will science ever fail?, *New Scientist* 1992, **8 August**, 32–5.

Atkins, P., *On Being: A Scientist's Exploration of the Great Questions of Existence*. Oxford: Oxford University Press, 2011.

Barker, M., *Creation*. London: T&T Clarke, 2010.

Barzun, J., *Science the Glorious Entertainment*. New York: Harper and Row, 1964.

Beale, G.K., *The Book of Revelation*. New International Greek Testament Commentary. Grand Rapids, MI: Eerdmans, 1998.

Bender, T., From academic knowledge to democratic knowledge. In *Values in Higher Education* S. Robinson and C. Katulushieds (eds.). Cardiff: Aureus Publishing, 2005.

Berger, P.L. (ed.), *The Desecularization of the World: Resurgent Religion and World Politics*, Washington, DC: Ethics and Public Policy Centre and Eerdmans, 1999.

Braque, G., *Le Jour et la Nuit: Cahiers 1917–52*. Paris: Gallimard, 1988.

Brooks, M., *The Secret Anarchy of Science*. London: Profile Books, 2012.

Brown, W.P., *The Seven Pillars of Creation: The Bible, Science, and the Ecology of Wonder*. New York: Oxford University Press, 2010.

Bryson, B., *A Short History of Nearly Everything*. London: Transworld, 2003.

Burke, D., Assessing risk: science or art?, *Science and Christian Belief* 2004, **16**, 27–44.

Burnett, C. (ed.), *Adelard of Bath, Conversations with his Nephew: On the Same and the Different, Questions on Natural Science, and On Birds*. Cambridge: Cambridge University Press, 1998.

Cantor, G., *Michael Faraday, Sandemanian and Scientist*. London: Macmillan, 1991.

Chancellor of the Duchy of Lancaster. *Realising our Potential: A Strategy for Science, Engineering and Technology*. Cm2250. London: HMSO, 1993.

Clines, D.J., *Word Biblical Commentary*, Vol. 18B. Nashville: Thomas Nelson, 2011.

Clines, D., *Job*, Vol. 3. Bellingham, WA: Thomas Nelson, 2013.

Coakley, S., *Sacrifice Regained: Evolution, Cooperation, and God*. Gifford Lecture Series 2012, University of Aberdeen. <http://www.abdn.ac.uk/gifford/about/> (Accessed 12 January 2014).

Collins, F., *The Language of God*. New York: Free Press, 2006.

Committee on the Conduct of Science, *On Being a Scientist*, 2nd ed. Washington, DC: National Academy Press, 1994.

Crombie, A.C., *Robert Grosseteste and the Origins of Experimental Science*. Oxford: Oxford University Press, 1953.

Davies, R.P.W., A. Aggeli, A.J. Beevers, N. Boden, L.M. Carrick, C.W.G. Fishwick, T.C.B. McLeish, I. Nyrkova and A.N. Semenov, Self-assembling β-sheet tape forming peptides, *Supramolecular Chemistry* 2006, **18**, 435–43.

Davies, S., P. Macnaghten and M. Kearnes (eds), *Reconfiguring Responsibility: Deepening Debate on Nanotechnology*. Durham: Durham University, 2009.

Dawkins, R., The emptiness of theology. *Free Inquiry magazine*, 1998, **18**, 2.

Dawkins, R., *The God Delusion*. London: Bantam, 2006.

Dawkins, R., *The Root of All Evil*. Channel 4, UK, 2006.

de Laplace, P.S., *A Philosophical Essay on Probabilities*. Transl. F.W. Truscott and F.L. Emory. New York: Dover Publications, 1951.

Deane-Drummond, C., *Creation through Wisdom: Theology and the New Biology*. Edinburgh: T&T Clarke, 2000.

Dennett, D.C., *Breaking the Spell: Religion as a Natural Phenomenon*. London: Allen Lane, 2006.

Dinkova-Brun, G., et al., *Dimensions of Colour: Robert Grosseteste's De Colore*. Durham: Institute of Medieval and Renaissance Studies, 2013.

Dunn, J., Romans 1–8. In *World Biblical Commentary*, Vol. 38A. Nashville: Thomas Nelson, 1988.

Dupuy, J.-P., The narratology of lay ethics, *Nanoethics* 2010, **4**, 153–70.

Durkheim, É., *The Elementary Forms of the Religious Life*. Transl. Carol Cosman. Oxford World Classics. Oxford: Oxford University Press, 1912.

Falk, D.R., *Coming to Peace with Science*. Downers Grove, IL: Inter-varsity Press, 2004.

Fanelli, D., How many scientists fabricate and falsify research? A systematic review and meta-analysis of survey data, *PLoS ONE* 2009, **4** (5), e5738.

Faraday, M., *Proceedings of the Royal Institution* 1829, **April–June**, 364.

Feynman, R.P., *The Feynman Lectures on Physics*, Vol. II. New York: Basic Books, 1964.

Feyerabend, P., *Against Method*. London: New Left Books, 1975.

Feyerabend, P., *Against Method*, 4th ed. New York: Verso Books, 2010.

Food and Agriculture Organization of the United Nations, *Statistical Yearbook 2013*. Rome: UN, 2013.

Geo, C. and S. Geo, *Mother Nature Becoming More Irritated As Even Geoengineering Scientists Admit They Might Be Making Things Worse . . . But They'll Keep Doing It, Anyway*. <http://www.geoengineeringwatch.org/> (Accessed 13 March 2013).

Gleick, J., *Chaos: Making a New Science*. London: Viking Penguin, 1987.

Gould, S.J., *Rocks of Ages: Science and Religion in the Fullness of Life*. New York: Ballantine Books, 2002.

Gregory of Nyssa, *On the Soul and the Resurrection*. Transl. C.P. Roth. New York: St. Vladimir's Seminary Press, 1993.

Habel, N.C., *The Book of Job*. London: SCM Press, 1985, p. 104.

Habel, N.C. and S. Wurst (eds.), *The Earth Story in Wisdom Traditions*, Sheffield: Sheffield Academic Press, 2001.

Habermas, J., Religion in the public sphere, *European Journal of Philosophy* 2006, **14**, 1–25.

Harrison, P., *The Fall of Man and the Foundations of Science*. Cambridge: Cambridge University Press, 2007.

Hedley Brooke, J., *Science and Religion: Some Historical Perspectives*. Cambridge: Cambridge University Press, 1991.

Heisenberg, W., *Physics and Philosophy*. Transl. from *Physik und Philosophie*. London: Allen and Unwin, 1959.

Hitchins, C., *God is not Great*. London: Atlantic Books, 2007.

Holmes, R., *The Age of Wonder: How the Romantic Generation Discovered the Beauty and Terror of Science*. London: Harper, 2009.

Iqbal, M., *The Reconstruction of Islamic Thought in Islam*. Oxford: Oxford University Press, 1934.

Kay, W.K., Male and female conceptualisations of science at the interface with religious education. In *Christian Theology and Religious Education* J. Astley and L.J. Francis (eds.). London: SPCK, 1995.

Kendall, C. and F. Wallis, *Bede: On the Nature of Things and On Times*. Liverpool: Liverpool University Press, 2010.

Kuhn, T., *The Structure of Scientific Revolutions*. Chicago: University of Chicago Press, 1962.

Lash, N., *The Beginning and End of 'Religion'*. Cambridge: Cambridge University Press, 1996, pp. 112–31.

Latour, B., 'It's development, stupid!' or: how to modernize modernization. In *Postenvironmentalism* J. Proctor (ed.). Cambridge, MA: MIT Press, 2008.

Latour, B., *Politics of Nature: How to Bring the Sciences into Democracy*. Transl. Catherine Porter. Cambridge, MA: Harvard University Press, 2004.

Lewis, C.S., *The Great Divorce*. New York: Macmillan, 1946.

Mabberley, D.J., *Jupiter Botanicus: Robert Brown of the British Museum*. London: British Museum (Natural History), 1985.

Main, I., Debate: Is the reliable prediction of individual earthquakes a realistic scientific goal?, *Nature* 1999, 25 February. http://www.nature.com/nature/debates/earthquake/equake_frameset.html

McEvilley, T., *The Shape of Ancient Thought: Comparative Studies in Greek and Indian Philosophies*. New York: Allwarth Press, 2002.

McLeish, T.C.B. and W.W. Graessley, The Doi–Edwards theory. In *Stealing the Gold: A Celebration of the Pioneering Physics of Sam Edwards*. Oxford: Clarendon Press, 2004.

McLeish, T.C.B. and W.C.K. Poon, How many cultures? 'Real presences' and the healing of the academy, *Interdisciplinary Science Reviews* 2001, **26**, 167–72.

Newberg, A., et al., *Why God Won't Go Away: Brain Science and the Biology of Belief*. New York: Random House, 2002.

Newsom, C., The book of Job. In *The New Interpreter's Bible*, Vol. 4, L.E. Keck et al. (eds.). Nashville: Abingdon, 1996.

O'Neill, O., *A Question of Trust: The BBC Reith Lectures 2002*, Cambridge: Cambridge University Press, 2002. http://www.bbc.co.uk/radio4/reith2002/

Paars, S., *Creation and Judgement: Creation Texts in Some Eighth Century Prophets*. Leiden: Brill, 2003.

Paine, T., *Age of Reason*, Part II, Section 21. Paris: Barrois, 1795.

Pais, A., *Subtle is the Lord: The Science and the Life of Albert Einstein*. Oxford: Oxford University Press, 1982.

Panti, C., Robert Grosseteste's *De luce*: a critical edition. In *Robert Grosseteste and his Intellectual Milieu: New Editions and Studies*, J. Flood, J.R. Ginther and J.W. Goering (eds.). Toronto: Pontifical Institute of Medieval Studies, 2013.

Parsons, K. (ed.), *The Science Wars: Debating Scientific Knowledge and Technology*. Amherst, NY: Prometheus Books, 2003.

Polanyi, M., *Personal Knowledge: Towards a Post-Critical Philosophy*. Chicago: University of Chicago Press, 1962.

Polkinghorne, J., *Science and Christian Belief*. London: SPCK, 1994.

Polkinghorne, J., *Science and Theology*. London: SPCK, 1998.

Polkinghorne, J., *Exploring Reality: The Intertwining of Science & Religion*. London: SPCK, 2005.

Popper, K., *The Logic of Scientific Discovery*. London: Routledge, 2002.

Ram, A., The making of optical glass in India: its lessons for industrial development, *Proceedings of the National Institute of Sciences of India* 1961, **27**, 564–5.

Reinhardt, O. and D.R. Oldroyd, Kant's theory of earthquakes and volcanic action, *Annals of Science* 1983, **40**, 247–72.

Robertson, D., The book of Job: a literary study. *Soundings* 1973, **56**, 446–68.

Sacks, J., *The Great Partnership: God, Science and the Search for Meaning*. London: Hodder and Stoughton, 2011.

Secretary of State for Energy and Climate Change, *Government Response to the House of Commons Science and Technology 8th Report of Session 2009–10: The Disclosure of Climate Data from the Climatic Research Unit at the University of East Anglia*. London: The Stationery Office, 2010. <http://www.official-documents.gov.uk/document/cm79/7934/7934.pdf>.

Silvas, A.M., *Macrina the Younger. Philosopher of God*. Turnhout: Brepols, 2008.

Smolin, L., *The Trouble with Physics: The Rise of String Theory, The Fall of a Science, and What Comes Next*. New York: Mariner Books, 2007.

Snow, C.P., *The Two Cultures*. Cambridge: Cambridge University Press, 1998.

Sokal, A., *Impostures Intellectuelles*. Paris: Éditions Odile Jacob, 1997.

Soloveitchik, J.B., *Halakhic Man*. Philadelphia, PA: Jewish Publication Society of America, 1984.

Steiner, G., *Real Presences*. London: Faber, 1989.

Steiner, G., *The University Festival Lecture: 'A Festival Overture'*. Edinburgh: University of Edinburgh, 1996.

Stenger, V.J., *The New Atheists: Taking a Stand for Science and Reason*. New York, NY: Prometheus, 2009.

Stokes, P.D., *Creativity from Constraint: The Psychology of Breakthrough*. New York: Springer, 2006.

Tilby, A., *Science and the Soul*. London: SPCK, 1992.

Tolkien, J.R.R., *On Fairy-Stories*. The Tolkien Reader. New York: Ballantine Books, 1966.

Turney, J., *Frankenstein's Footsteps*. New Haven, CT: Yale University Press, 1998.

Turney, J., *Frankenstein's Footsteps*, 2nd ed. New Haven, CT: Yale University Press, 2000.

United Nations General Assembly, Fifty-fifth session agenda item 60(b), A/RES/55/2. UN, 2000. <http://www.un.org/millennium/declaration/ares552e.pdf>

Vanel, L., et al., Memories in sand: experimental tests of construction history on stress distributions under sandpiles, *Physical Review E* 1999, **60**, R5040–3.

Vanstone, W.H., *The Stature of Waiting*. London: Longman and Todd, 1982.

Voltaire, *Candide*. London: Penguin, 1997 (first published 1759).

Weinberg, S., Can science explain everything? Anything? In *The Best American Science Writing* M. Ridley (ed.). New York: HarperCollins, 2002.

Westfall, R.S., *Never at Rest: A Biography of Isaac Newton*. Cambridge: Cambridge University Press, 1980.

Wilcock, M., *The Message of Revelation*. Nottingham: Inter-varsity Press, 1991.

Williams, R., Faith in the university. In *Values in Higher Education* S. Robinson and C. Katulushi (eds.). Cardiff: Aureus Publishing, 2005.

Wittmer, J.P., et al., An explanation for the central stress minimum in sand piles, *Nature* 1996, **382**, 336–8.

Wolfers, D., Science in the book of Job, *Jewish Bible Quarterly* 1990, **19**, 18–21.

Wright, N.T., *The New Testament and the People of God*. Christian Origins and the Question of God, Vol. 1. London: SPCK, 1992.

Wright, N.T., *The Resurrection of the Son of God*. Christian Origins and the Question of God, Vol. 3. London: SPCK, 2003.

Wright, N.T., *Surprised by Hope*. New York: HarperCollins, 2008.

Zinser, T.J., *Examination of issues related to internet posting of emails from Climatic Research Unit*, 18 February 2011. Washington, DC: Office of the Inspector General of the United States Department of Commerce. http://www.oig.doc.gov/OIGPublications/2011.02.18-IG-to-Inhofe.pdf

Zuriguel, I. and T. Mullins, The role of particle shape on the stress distribution in a sandpile, *Proceedings of the Royal Society A* 2011, **464**, 99–116.

Index of Biblical Passages

Index

Made in the USA
San Bernardino, CA
09 December 2016